NEW VISIONS IN PERFORMANCE

Innovations in art and design

Series editor:
Colin Beardon, University of Waikato, New Zealand

New Visions in Performance.
The Impact of Digital Technologies

Edited by

Gavin Carver
University of Kent, UK

and

Colin Beardon
University of Waikato, New Zealand

SWETS & ZEITLINGER
PUBLISHERS

LISSE   ABINGDON   EXTON (PA)   TOKYO

Library of Congress Cataloging-in-Publication Data

A Catalogue record for the book is available from the Library of Congress

Printed in the Netherlands by Krips The Print Force, Meppel

Published by: Swets & Zeitlinger Publishers, www.szp.swets.nl

ISBN 90 265 1966 4
ISSN 1570-7393

Contents

1 Introduction

Gavin Carver and Colin Beardon

The range of content

Digital technologies have, until fairly recently, been principally employed in theatre as an invisible (behind the scenes) tool, contributing to the efficient or spectacular staging of performance. While of course it is the case that media such as film, television and video have been integrated into performance throughout the 20th century, particularly the last 40 years, the creative utilisation of digital processing technology (often coupled with image or sound based media) is far more recent. The performances and projects described and analysed in this book have fully integrated digital technologies in the conceptual, creative and performance processes, opening-up new possibilities of performance and its relationship to time, space and the body. This triumvirate, along with the associated, often interrelated notions of interactivity and liveness form the crux of the new visions in our title. For these authors this is not merely playing with a new toy (although understandably many of the projects illustrate enthusiastic and playful investigation), nor is it simply a matter of efficiency and expediency; rather these projects, and their analyses, provide a rich vein of new theoretical and creative concerns.

The chapters in this book have been written by performance makers and academics (often one and the same), and all provide case studies of the deployment of new technologies in, or around, performance. Some are written from the position of the critical observer, some from the position of an involved artist, but in all cases the discussion is around the practice and its theoretical implications. We have intentionally sought to ensure a wide range of work is represented, the chapters describe projects from a wide spectrum of performance forms, contexts and modes of integrating digital technologies into the process and performance. Thus, the projects described range from identifiable theatrical performances (in that they include distinct bodies of performers and audience, brought together in a theatre space), through interactive projects where 'participants' (hybrid performer / audience) drive the story (often in part, or in totality projected as if it were a film), through to video installation work. Some projects generally develop from the conventions of drama, several emerge from dance and physical performance, and others have a genealogical connection with film and computer games. The work described is produced by or in a range of settings: large national institutional theatres, fringe or experimental venues and university departments. Together they illustrate the range of possibilities rather than create an impression of a particular flavour or style of work: new media no longer (if it ever did) suggests solely a practice centred on cameras, monitors and postmodern fragmentation and cynicism. While all the work described here employs digital media very little of it is actually about digital media, or mass communication in general. Inevitably, the focus of many articles is on the development and use of the technology, but many of the final performances described do not engage, overtly or directly, with the social and cultural ramifications of a digital age. Of course, such concerns are implicit, as much as the effect of the technological revolutions of 19th Century Britain are implicit in Victorian Melodrama while only a handful of the plays directly tackled the subject.

Themes of the book

The breadth of examples covered in the book makes it unwise to attempt to extrapolate a unifying theory of new technology in performance from the examples cited. Firstly, there are a range of performance forms which have their own poetics and critical theories, and which are making use of digital systems to advance, enhance or test their practices. It would be foolish to imagine that contemporary dance and interactive film have sufficient in common (other than their use of similar media) to create a new coherent common discourse. Secondly there are, inevitably, practices not included here and studies of these would provide further insights into the form. However what we can do is to see what are the central, or common themes of this snapshot of performance in the 21st century.

Reading the chapters included here it becomes clear that while describing widely differing projects a number of common concerns emerge. Simplified, these are: the virtuality and fluidity of space and time, and the potential for alternative realities, spaces and narratives; interactivity and the active audience / participant; the role of the body (and its double) in technologically enhanced or mediated performance and the ensuing questions around identity and presence; the ability of performance to extend itself beyond the circumscribed moment and place of its enunciation; and the 'problem' of liveness in multimedia work—issues that inevitably (predictably perhaps) question taxonomies in performance. Associated issues around the cultural and pedagogical politics of technology also arise in the works, tangentially in many, but directly in some.

In the following chapters these complex and rich veins of exploration braid together making a thematic separation of the chapters undesirable if not impossible, more importantly each piece stands alone in its description of a unique project that was designed (and written about) with its own integrity, not tailored to fit a constructed meta-theory. Nonetheless certain themes inevitably come to the fore in each of the articles. Mark Coniglio, as his title suggests, considers (and celebrates) the importance of a direct feedback between the movements of a performer and the digitally controlled environment, a theme pursued by Jools Gilson-Ellis and Richard Povall, and featured in a number of other chapters: Dove, Callesen et al. For Steve Dixon the ability of digital systems to create doubles of the performer is the central concern, and the chapter considers the ontological and theatrical implications of these doubles and by extension the 'body' of the performer; doubles emerge in other works, notably projects cited by Kjell Yngve Petersen and Toni Dove. Petersen and Smart particularly consider the manipulation of space and time in performance; for Petersen the interest is principally in real time disruption, while in Smart's analysis of McGregor's dance piece *Nemesis* the focus is on the choreographed and designed presentation of the body in a problematised framework of space and time. Katrine Nilsen, Jorgen Callesen and Marika Kajo clearly identify and critique their work as research rather than solely creative output, and articulate the process of designing experiments that investigate the relationship between an environment controlled by an interactive system and the participation of an active audience. Somewhere between structured play and participatory performance the projects not only address issues arising for flexible interactive performance structures, where time and space are responsive and plastic, but also seek to understand how these projects impact on traditional definitions of roles in performance.

Jeff Burke and Jared Stein also involve the audience as active participant in their work, reflecting on how the input (and mediated output) of data by and from the audience can become a central part of the performance text. In analysing their *Iliad* project they also question the common applications of interactive and flexible, reader defined narratives. Toni Dove analyses two of her projects where the audience control the unfolding of the filmed narrative, almost as if from within the story; the chapter not only addresses issues of time, interactivity and active audiences but also raises question about presence, liveness and the intermediality of such work. Similarly Michael Nitsche and Maureen Thomas utilise film reference, tropes and modes to create a 'live' performance, and in doing so bring the readers' attention to the intersection and space between film, theatre and computer games; the use of networked technologies makes their performance 'remote', but it is nonetheless live and interactive. The essay also indicates the pedagogical usefulness of virtual systems and touches on issues relating to the relationship between avatar and performer.

Harris' performances are also remote, offering an artwork that is derived from a costumed dancer, but where, in the final showing the body has been removed. Liveness, presence, the body are all concerns of this chapter which also articulates the personal responses of the artist to the developing processes and technologies.

Finally Christie Carson extends her study beyond the moment and place of performance, exploring how digital technologies can effect performance live through their impact on the culture and framing of theatre, ultimately indicating (touched on by several other chapters) that the advantages of new communications technologies lies in their ability to tailor accessible information, this changing context of performance may well be as significant as the changing modes of performance.

Such is the richness of these enquiries that we (the editors) felt that the material could be better explored in a summarising chapter, placed after the main contributions. So rather than attempt to develop any more thematic material before the reader has read the works, this 'introduction' continues in Chapter 13…

2 The importance of being interactive

Mark Coniglio

My collaborator Dawn Stoppiello and I created our dance theatre company *Troika Ranch* in 1994, our purpose being to create dynamic, challenging artworks that fused traditional elements of dance, music and theatre with interactive digital media. We believed that by directly linking the actions of a performer to the sound and imagery that accompanied them, we would be led to new modes of creation and performance and, eventually, to a new form of live art work. While we cannot yet claim to have reached this latter, rather lofty, goal we have firmly established our views about interactive performance and its importance to the performer and audience. Presenting these views is the subject of this chapter, but before I go on, I think it is worth answering a simple question: why would one want to create such artworks in the first place?

Live media / dead media

The answer begins with the a device I created in 1989 called *MidiDancer*, a sensory system that uses flexion sensors to wirelessly report the position of a dancer's joints to a computer. Software interprets the movement information, which can in turn manipulate digital media in a numerous ways: initiating the playback of musical notes or phrases, manipulating live or pre-recorded video imagery, and controlling theatrical lighting are just three of the possibilities. But *MidiDancer* was not so much an answer as the beginning of the question posed above. The personal computer technology of that time had made it possible for us to use dancer's movement to generate a musical accompaniment. In other words, we did it because it had become possible to do so. We had no true notion as to why it was essential to the aesthetic expression.

It was not until 1996 that a clear answer to the question of 'why' presented itself to me. Around that time, Dawn and I had a residency at STEIM (Studio for Electro Instrumental Music) in Amsterdam. During the first three days of that residency, four separate individuals asked in passing, "Has Jorgen taken you to his room yet?" Now Jorgen, whose specialty was electronic fabrication, was one of several talented engineers at STEIM who helped the artists in residence to realise

Figure 1. The author performing in *The Chemical Wedding of Christian Rosenkreutz*, June 2000.

their projects. With my curiosity piqued, I finally sought him out and asked rather sheepishly, "Jorgen, can I see your room?" After nodding in the affirmative, he led me up four very steep and narrow staircases to the attic of the building. There, we came to a single white door, which he opened with some ceremony. Inside was revealed the most elaborate and extensive collection of 1960s and 1970s era analogue synthesizers I had ever seen assembled in one location. After showing me around, he proceeded to play me several examples of his music using a single analogue sequencer to control every instrument in the room. Now, this being 1996, and at this point being completely ensconced in using digital technology myself, I was compelled to ask, "Is anything in this room digital?" He shot me a glance and replied, "Oh no!" When asked "Why not?" he replied quite seriously "Because it is always the same."

In that moment I realised that what we love about digital media was precisely what made it inappropriate for use in a live performance—it is indeed always the same. Digital media is wonderful because it can be endlessly duplicated and/or presented without fear of the tiniest change or degradation. But, it is this very quality (the media's 'deadness') that is antithetical to the fluid and ever changing nature of live performance. Each time a work is performed, any number of factors can significantly change how it is realised in that moment—perhaps most significant being the interplay between the skill and temperament of the performers and the attitude and engagement of the audience. The downfall of digitally recorded media is that it dampens this essential fluidity by preventing the performers from changing the character of the material from moment to moment.

Figure 2. The Company in *Reine Rien* for four *MidiDancers* and interactive media, September 2001.

Organic -> electronic

I often witness the tension between recorded media and live performance when I attend performances in one of New York City's small, alternative modern dance venues. It is almost a given in these situations that the dancers will perform to music pre-recorded on a compact disc. On any given night, these performers have the potential to give the performance of a lifetime, given the right combination of skill, an understanding of their instrument (i.e. their body) in relation to the material that they are to perform, and an awareness of the nebulous (but, as any performer will acknowledge, real) feedback loop between performer and audience. But, when the performers attempt to nuance a gesture or phrase in response to the aforementioned relationships, an unrelenting and unaware companion—the digitally recorded music with which they perform—thwarts them. In this situation, they cannot hold a spectacular balance because if they did, the music would race on ahead of them and a subsequent phrase of the dance would suffer as they attempt to catch up.

So, the answer to my own question is, I provide interactive control to the performers as a way of imposing the chaos of the organic on to the fixed nature of the electronic, ensuring that the digital materials remain as fluid and alive as the performers themselves. There are two important implications that arise from this approach, namely:

1. we must give the performers latitude to improvise if they are to take advantage of such interactivity; and
2. the audience must have some understanding of the interaction to complete the loop between audience and performer.

I want to examine these points in some detail, using musical models of performance that are rendered more expressive because of their real-time interaction.

In terms of providing interactive control, and the importance of improvisational decision making, let us consider the classical orchestra. The music played by the orchestra (the 'media') has been precisely notated well in advance of the performance. Yet, relying on the same skill, awareness and feedback from the audience cited above, it is the conductor who will determine the music's timing and dynamics from moment to moment, and thus its final realisation. (This model seems particularly appropriate when applied to interactive dance performance, as both rely on gesture as the means of interactive control.) The conductor of the classical orchestra does not traditionally reformulate the music in a manner so radical as to change the nature of the piece, but it is certainly possible. And, because the conductor is at some level improvising, it is possible that such a reformulation could happen at any time.

It could be argued that by changing only two parameters (again, timing and dynamics) the conductor cannot fundamentally change a musical work. The converse was proved to me by an exercise I witnessed as a student. My teacher Morton Subotnick gave the dozen or so composers in my class the first two pages of Pierre Boulez's *Piano Sonata No. 1* with the following instructions: we were to change the dynamics and octave transposition[1] of the notes freely, but no other parameter was to be changed. The following week each of our manipulated versions was performed for us, but without identifying who had done the work. To my amazement, I found that I could detect who had 'composed' each and every version because I was intimately familiar with the style of my colleagues. Even with only two parameters open to change, the results were absolutely personal. It follows then that a skilled interactive performer would be equally able to impose their own, quite personal, interpretation on the pre-composed media materials under their control—even if the number of parameters they can manipulate are limited. This belief has become a core strategy in my approach to creating live interactive art works.

It is worth noting that, while we could manipulate only two parameters in the exercise, we were allowed to do so 'freely'—without limitation. As applied to live interaction this tells us that, while the number of parameters that a performer can manipulate might be limited, the range of those manipulations must be profound enough to allow the performer to place his or her personal interpretive stamp on the material.

Essential improvisation

Another musical model to consider is jazz, as it speaks clearly to the importance of improvisation and to the audience's understanding of that process. Take the specific example of a jazz pianist. It may seem obvious, but because of historical or personal experience we know that when a finger is placed against a key on the piano, and the key is pressed down, a musical note is produced. This taken-for-granted understanding allows us to 'know' that the pianist is playing his instrument as we

Figure 3. Danielle Goldman and Sandra Tillett in *Future of Memory*, February 2003.

watch him perform. (Indeed, quite a scandal ensues if the audience discovers this relationship has been faked—see Milli Vanilli, circa 1990.) If we know the jazz form, we come with the understanding and expectation that the music will be invented in the moment of performance. So, we experience a certain kind of thrill as we watch a performer instantaneously organise and artfully play musical materials before our eyes. The real-time nature of the music's creation is so integral that it is part and parcel of its meaning. To not understand this is to not be able to fully appreciate the art form.

It would seem that quite a similar appreciation should be possible for interactive performances, but in fact there are several obstacles to overcome. When presenting such works, we cannot *a priori* rely on the expectations and understanding the jazz audience brings to a performance. This is because:

1. the audience may not be aware that there is some level of improvisation occurring, and
2. because the audience has no prior understanding of the 'instrument' with which the performer controls that manipulation.

Regarding these two points, consider my company *Troika Ranch*. We create dance theatre works that are most often presented on a proscenium stage, a setting that historically features work that is composed in advance. So, while we do give our performers a fairly broad range of improvisational latitude when performing, the audience will generally assume that what they are seeing is not improvised at all if only because of the setting in which the work takes place. This problem of perception is exacerbated by the audience's lack of experience with the instruments

Figure 4. Real-time video capture and manipulation in *Future of Memory*, February 2003.

being used to manipulate the media. Our dancers wear wireless sensors on their bodies (*MidiDancer*) that allow them to manipulate digital media in real time. These interactive instruments are quite similar to their musical counterparts in that they translate gesture into another form, so one might assume that the audience could easily understand their function. In fact, because these instruments are new, unique, and unfamiliar, the audience has no historical or personal experience with them. So, in practice, it is quite difficult for the audience to perceive that a *Troika Ranch* performance is in fact quite similar to the jazz performance described above.

This is not to say that an interactive work cannot be appreciated at face value. In our example of the jazz pianist, even those who do not understand the inner workings of a piano or that the performer is improvising can appreciate the musical result. So too it should be with interactive performance. But, an audience's understanding that the performer has a virtuosic command of his or her instrument and that he or she is creating something new in the moment of performance adds yet another layer of 'liveness' to the experience, which I would argue is a core rationale for adding interaction to the mix in the first place.

Fundamental increments

I want to entertain a worthwhile assertion by one colleague of mine. Namely, that my use of the musical models described above is reactionary because it imposes a kind of tunnel vision, inhibiting the development of truly new grammars/strategies for the creation and realisation of new performance. In my view, this argument is flawed because it presupposes that the use of sensory technologies and digital media offers some kind of radical shift in the nature of live performance itself.

Countering this argument fully is beyond the scope of this essay, but let me address it briefly by considering what I think we can agree was a revolutionary technological innovation: photography. The ability to immediately capture and reproduce an image changed the public's experience of the world almost overnight. Take as an example the early horrific images of soldiers strewn over the battlefield during the U.S. Civil War. These widely published photographs brought an immediacy of experience to the viewer that was inconceivable previously and shaped public opinion of that war. If the integration of cutting edge technology in performance was similarly radical in nature, would not its impact be as immediate and unstoppable as that of photography?

There is a vague feeling in our time that any new and sufficiently unfamiliar technology holds the promise of radically altering the fabric of our society. This belief is in part factual, based on the transformational technological innovations of the late 19th and early 20th centuries. It is also based on highly suspect marketing claims that reached their saturation apex at the height of the Internet boom. And, in the field of live technology and performance, artists have often enough made public claim to the potential for revolutionary new forms. (I must plead guilty to doing just that during the early days of the *MidiDancer*—not that I didn't believe it to be true at the time.) I would argue, however, that the ramifications of new media and sensory technology in performance are much more likely to be incremental than fundamental.

The software program *Photoshop* provides a useful example. Its central metaphor of painting on a canvas is rooted in a tradition that everyone understands. The ability to save multiple versions, the notion of 'undo', and the introduction of algorithmic processes that can be applied to the image (i.e. filters) do change the working process in important ways. But, while we can agree that skilled artists have used this tool to make images that are arresting, stunning or beautiful, the nature of the image itself has not been changed. This lies in contrast to photography, which depicted the world with a sense of reality (implying both truth and objectivity) that had never been experienced before—it redefined image.

Now, my colleague might reasonably argue that the reason that *Photoshop* did not redefine the image was because its metaphor was based on existing models of creating imagery. I would counter that it was not possible to go beyond the existing models because *Photoshop* simply did not alter the essential notion of image. The impact of new technology and interactivity on live performance is far more akin to that of *Photoshop* than it is to photography, and so it seems to me that applying existing models to technologically enhanced performance is valid and useful.

Regardless, I do think that the use of interactivity in live performance is essential. Live performance is perhaps the most inefficient of contemporary art forms, because you cannot do with it what you can with digitally stored artworks: duplicate and inexpensively deliver it to a large audience. But, it is specifically the ineffable quality of liveness that draws me to create and attend performance. By using new technology to allow our performers to become real-time creators, and by asking our audience to be present to their on-the-fly artistry, we ensure that each performance of a work is absolutely unrepeatable, which may be the boldest move of all.

Note

[1] An 'A' could not become a 'B-flat', but instead the note could be transposed up or down by one or more octaves.

3 The digital double

Steve Dixon

Introduction: theatre and its digital double

The notion of the double has been a potent concept in performance since the publication of Antonin Artaud's *The theatre and its double* in 1938. The metaphor is now becoming concrete and actuated in the theory and practice of performance which foregrounds the use of computer technologies; what I term 'digital performance'. Artaud's notion of theatre's double is a primitivist and spiritualised vision of a sacred, transformational and transcendental theatre. Early on in *The theatre and its double*, Artaud discusses the totems of Mexican culture: magically-invested rocks, animals and objects which can excite dormant powers in those who worship or meditate upon them. In a typically abrupt, jagged thought-change (now ubiquitous in computational hypermedia), in the space of a single sentence this totem suddenly becomes an effigy, a double and a shadow: "All true effigies have a double, a shadowed self" (Artaud 1974 5). For Artaud, the double of theatre is its true and magical self, stirring other dark and potent shadows which rail against a fossilised, shadowless culture "as empty as it is saccharined" (Artaud 1974 34). Discourses on cyberculture now reinscribe this Artaudian dialectic, where a romantic utopianism hailing spiritualised virtual realities is pitted against a dystopian scepticism, which attacks the soulless, alienated and schizoid nature of digital irreality.

Artaud's incendiary theatrical writings conjure images, previously considered impossible to stage, which are now being realised using the capabilities of the computer. His belief that actors should be "like those tortured at the stake, signalling through the flames" (Artaud 1974 6) has been echoed in works such as 4D Art's dance-theatre production *Anima* (2002) and Marcel-Lí. Anthúnez Roca's one-man show, *Afasia* (1998), which features a vivid computer-manipulated projection of his rotating, flame-licked body. Images of performers' digital doubles are telematically transmitted to different locations where they dance or interact with distant partners in real-time; and in 'single-space' theatre events, performers' projected body doubles now commonly play dialogues and perform duets with their live counterparts. The digital double projects itself online and on stage to take numerous forms, from the textual characterisations of role-playing MUDs and

Figure 1. Ruth Gibson dances with her digital double in Igloo's *Viking Shoppers* (2000).

MOOs to the graphical avatars of virtual worlds; from the theatrical depictions of cyborgic alter-egos to the parthenogenic creations of artists' substitute-selves in the form of anthropomorphic robots.

In his discussion of theatre as alchemy, Artaud became the first person to coin the term 'virtual reality'. Linking the chimeric nature of both theatre and alchemical symbols he describes how "theatre's virtual reality develops… [on the] dreamlike level on which alchemist signs are evolved" (Artaud 1974 35). For Artaud, since alchemical signs are "like a mental Double of an act effective on the level of real matter alone, theatre ought to be considered as the Double, not of this immediate, everyday reality … but another, deadlier archetypal reality" (Artaud 1974 34). A doubled reality or 'virtuality' becomes the fulcrum for Artaud's theatre of cruelty which, like the Balinese dance he witnessed in 1931, enacted "a virtuality whose double produced this intense scenic poetry, this many-hued spatial language" (Artaud 1974 46).

The idea of the body and its double pervades digital performance, and relates to the shadow figure of the doppelgänger, Freudian notions of the uncanny and the subconscious Id, and Jacques Lacan's concept of the mirror stage and the *corps morcele* (the body in pieces). Lacan's theory of the mirror stage emphasises the misconceived, 'fictitious invention' of identity and the ego. This is conceived through the narcissistic lure of the apparently whole yet ultimately phantasmic projection of the subject's body double in the frame of the mirror "by which the subject anticipates in a mirage the maturation of his power" (Lacan 1977 2).

Identity is thus falsely constructed as a unity, whilst "by identifying with the imaginary mirage of the whole body, the 'real' of the fragmented body is repressed" (Herzogenrath 2002). Sigmund Freud presents images of the *corps morcele* ("dismembered limbs, a severed head, a hand cut off at the wrist" (Freud 1985 366)) as defining symbols of *The uncanny* (*Das unheimlich*), as well as robotic doubles such as the seductive mechanical doll, Olympia, in *The tales of Hoffmann* (Hoffmann 1982). Freud's notion of the uncanny (*unheimlich*) concerns the emergence of a dark self or 'other' in the midst of the familiar and normal: "the unheimlich is what was once heimisch, home-like, familiar; the prefix 'un' [un-] is the token of repression" (Freud 1985 368). When repressive barriers to the subconscious are pricked or come down, the uncanny may emerge to create a doubled reality where the familiar becomes frighteningly unfamiliar, "laying bare … hidden forces" (Freud 1985 366).

Jacques Derrida returns to Freud's *The uncanny* time and time again, writing in his essay *The double session* of Freud's fantastic and compelling "paradoxes of the double and of repetition, the blurring of boundary lines between 'imagination' and 'reality', between the 'symbol' and the 'thing it symbolises'" (Derrida 1981 220). For Martin Heidegger, the uncanny is a fundamental condition of existential being, an anxious and fearful sense of separation from reality, both "tranquillised and familiar" where the feeling of "not-being-at-home [das Nicht-zuhause-sein]" is a "primordial phenomenon" (Heidegger 1962 233–34). Nicholas Royle's rich and exhaustive study, *The uncanny* (2003), traces the notion through the writings of Kant, Marx, Nietzsche, Freud, Heidegger, Wittgenstein and Derrida, and goes on to explore its relationship to Brecht's alienation-effect, to genetic engineering and cloning, and in a chapter where he antagonistically discusses Nicholas Royle's work in the third person, to the concept of the double. He also notes the value of the concept of the uncanny in understanding computer technologies:

> As a theory of the ghostly (the ghostliness of machines but also of feelings, concepts and beliefs), the uncanny is as much concerned with the question of computers and 'new technology' as it is with questions of religion. Spectrally affective and conceptual, demanding rationalization yet uncertainly exceeding or falling short of it, the uncanny offers new ways of thinking about the contemporary 'return to the religious' … as well as about the strangeness of 'programming' in general.
> (Royle 2003 23–24)

The double has an ancient and global lineage within religious, occult and folkloric traditions. The digital double relates to ancient notions of magic not only by virtue of the 'conjuring' of an alternate and simultaneous second body for the performing subject, but also in relation to the ancient laws of 'imitative' and 'homeopathic' magic. In his account of death and resurrection dramas and rituals, Sir James Frazer argues that these ceremonies "were in substance a dramatic representation of the natural processes which they wished to facilitate; for it is a familiar tenet of magic that you can produce any desired effect by merely imitating it" (Frazer 1925 324). In recent performance practice the double as a digital image replicating its human referent has been used to produce a range of different forms of imitation and representation which reflect upon the changing nature and understanding of the body and self, spirit, technology, and theatre.

In analysing the different manifestations currently being explored in performance I have identified four types or categories of the digital double which explore distinct representations and themes. These trace the double through the forms of a reflection, an alter-ego, a spiritual emanation, and a manipulable mannequin. Whilst these categories are useful in examining what I consider to be distinct aesthetic, philosophical or paradigmatic issues, it is nonetheless acknowledged that the double is by nature a mysterious and capricious figure which may sometimes challenge and traverse neat boundaries.

The double as reflection

Liquid Views (Monika Fleischmann/ Wolfgang Strauss/ Christian-A. Bohn 1993) epitomises the double as a narcissistic mirror reflection of Lacanian misrecognition ('meconnaissance'); the body appearing as "lines of 'fragilization' that define the anatomy of phantasy" (Lacan 1977 5). The gallery installation places the user in the role of Narcissus, peering into water to study her own reflection. The user/spectator becomes not only an interactive participant but also the primary subject and 'performer', since her digital double is also projected onto a large wall within the gallery space, to be watched by other visitors. Bending over a horizontal computer monitor, she sees a well or spring of 'virtual' water: rippling patterns of graphics programmed using custom software to imitate natural aquatic movement and wave shapes. A miniature video camera records the image of her face peering down, and computer software blends this picture with the moving water. A touch-screen interface enables the user to disturb and affect the water patterns, creating wave effects to blur the 'reflection', or to make it disappear in a swirling whirlpool.

This technological reworking of the Narcissus myth lends it a new and distinctly modern resonance. Whilst we appear to look at our own reflection, it is not a natural reflection, but a video copy—an electronic simulation transmitted through lenses, chips and cables, then reproduced as coloured pixels. The water is wholly synthetic, a series of algorithms prompting computer graphics. Narcissus sought the sublime through meditation on his own natural beauty, through natural reflective matter—water. In *Liquid Views*, a new metaphor and myth is etched out, and a highly resonant, contemporary image slowly crystallises. We are now in a period where we crane our heads and peer down to glimpse a new, technological sublime, which we find equally fascinating and hypnotic as the natural one. The image we peer at places ourselves firmly within, and in control of, a pulsating digital world simultaneously reflecting and synthetically replicating nature, and our own bodies. Human vanity is replaced by the new technological vanity: our faith in the transformational power of computer technology—the power of the virtual over the real. The installation is a simulacral mandala for meditation on the double as a technological reflection, where we gaze fixedly into, and back at, our new electronic selves.

Fascination with the double as a technological reflection becomes an obsession for the protagonist in Blast Theory's multimedia theatre performance *10 Backwards* (1999). The production gently and hypnotically explores notions of time and materiality through skilful interaction between the live 'play', pre-recorded video imagery, and a live onstage camera. The central character, Niki, videotapes herself compulsively, and the piece is punctuated with footage from her 'video diaries' where she speculates and fantasises about her future.

One of the production's most original and memorable sequences involves Niki using her digital video camera to record herself eating breakfast cereal. She eats exaggeratedly, repeatedly glancing at the camera and occasionally making other distinct movements: scratching her nose, adjusting her hair, picking cereal from her teeth with her fingers. Running against this live action, her recorded voice-over explains that her recording process is part of a strategy to re-orientate her mind and body in order to be able to travel through time. The close analysis and exact replication of her past actions, such as eating breakfast, can affect the temporal co-ordinates linking past, present and future.

She finishes eating, and begins to replay the footage she has just recorded, using a remote control 'jog-shuttle' mechanism to play the recording at different speeds, including frame-by-frame, forwards or backwards. The video plays on screens at each end of the traverse stage. As she slowly shuttles through the digital tape, she studies the screen image and synchronously re-enacts and mirrors every detail, action by action, grimace by grimace, chew by chew. She frequently holds the pause button, her live face and body frozen exactly as on the screens. She plays segments backwards, and once again her staccato live movements exactly mirror the screen footage as she adjusts her hair in reverse and brings a spoonful of cereal out of her mouth, and back into the bowl. The close-up video image on the screen

Figure 2. Niki Woods attempts to relive and replicate every facial detail of her breakfast-eating ritual with the help of a video 'jog-shuttle' device (in her left hand) in Blast Theory's *10 Backwards* (1999).

behind Niki is five times her corporeal size, rendering the mirrored interaction between the masticating live performer and her recorded self comic and grotesque. The voice-over continues throughout:

> Preparation number three, I was told, involves learning to love the jog-shuttle…
> Use it, abuse it, go as far as to break it, but run your everyday actions backwards,
> forwards, slowly, quickly. Don't forget the pause button. Cut your every move into
> a thousand pieces. Reassemble them as plasticine, twist them, stretch them, fold
> them in on themselves until you can regurgitate the breakfast before last. (Pause)
> By the fourth day of preparation, I'd learnt to hate my well-known brand of
> breakfast cereal.

Niki thus copies, analyses, replays, reverses and freezes the tiniest details of her breakfast meal. It is a humorous, strange and powerful theatrical ritual, where performance and technology truly and effectively meet. The live performer and her digital double exactly reflect and replicate one another within a context which seems at once personal but theatrical, banal yet profound. The universal act of eating is replayed as a slow, intense facial dance, a mirror-play duet between the live performer and her digital reflection. As we watch, our own facial muscles twitch empathetically, hardly able to resist joining in. The sequence synchronises Nikki to her past, and the audience to the apparently simultaneous 'presents' of her live and projected/recorded form. Niki uses digital video as an instrument for self-analysis, but for her the technological duplicate becomes the 'real' that she must painstakingly emulate and copy. The sequence's power and fascination lies in the fact that we see her gradually embodying and becoming 'one' with her technological reflection. Just as in *Liquid Views*, not only do the boundaries between the virtual and the actual become confused and unstable, but the digital reflection effectively effaces its live double to emerge as the dominant force.

Although all types of digital double can be conceptualised as some form of technological reflection of a live body, in my categorisations I am specifically defining the reflection double as a digital figure which mirrors the identical visual form and real-time movement of the performer or interactive user. In performances and installations featuring the reflection double, the performer or user is normally conscious of the presence of the double, as in *Liquid Views* and *10 Backwards*. In performances where the double co-exists with the live performer but is not directly watched and acknowledged by her, or where the double undertakes asynchronous activity, or presents another 'side' or visual embodiment of a character, we move into the domain of the digital double as an 'alter-ego'.

The double as alter-ego

The double as a performative alter-ego has been explored for almost a century in theatre productions which have incorporated film and video, and the audience's delight of recognition and differentiation between the live body and its mediatised double can be traced back to one of the earliest examples. In 1914, Winsor McCay toured the United States with *Gertie the Dinosaur*, in which he performed live on stage, issuing verbal and gestural commands to Gertie, a silent animated film character projected onto a movie screen at the back of the stage. Through McCay's precision timing, the female dinosaur animation appeared to respond instantly to his instructions by sitting up, rolling over, and performing tricks like a circus animal. The performance's climax involved McCay running off and disappearing behind the screen, whereupon he instantly reappeared running on the film as an animated stick figure, finally jumping onto Gertie's back to ride off into the distance (see Bell 2000 46).

Gilles Deleuze and Felix Guatarri's hugely influential *Mille plateaus* begins: "The two of us wrote *Anti-Oedipus* together. Since each of us was several, there was already quite a crowd" (Deleuze and Guatarri 1987 3). The double made digital flesh in contemporary multimedia performance commonly splits the several beings and underlying personalities of the performing subject and manifests this crowd. The digital double as an alter-ego is also able to encapsulate the different mystical 'becomings' at the centre of *Mille plateaus*: becoming-animal, becoming-intense, becoming-woman, becoming-stars. Hans Holzer describes how "the concept of the human double (or doubles) has remained a constant factor in folklore and tradition, particularly the belief that every human being is accompanied through life by two extensions of his personality, the one good and the other evil; the former luminous and the latter dark and menacing" (Holzer 1974 240). He suggests that this derives from "a primitive rationalisation into companion spirits of the reflected image in water and the ever-accompanying shadow". The shadow double of the alter-ego in digital performance is likewise an alternate, and invariably darker embodiment.

A visually striking example of the splitting of the alter-ego double from its human body occurs in 4D Art's performance *Anima* (2002). The company use a sophisticated projection system incorporating half-mirrored screens which are invisible to the naked eye, to create the illusion of three-dimensional human doubles appearing in the same physical space as the live performers, rather than on a projection screen. At one point a performer dances whilst projected after-images of her movements split away from her, hanging and moving in apparent 3D in the immediate space around her body. In another sequence, a male dancer's body appears to split into two separate bodies as his projected double tears itself entirely away from his live body and the two bodies perform a furiously physical and elaborate duet, continually circling and crossing one another's paths like aggressive fighters.

Such illusions of the apparitional double emerging from a live physical body relate to what Deleuze in *Cinema 2* defines as a 'crystal image' whereby "it is as if the real and the imaginary were running after each other, as if each were reflected in each other, around a point of indiscernibility" (quoted in Ryan 2001 33). But whilst Deleuze's notion of the crystal image offers a poetic metaphor for the digital double, his subsequent meditation on the totality of this indiscernibility stretches the point too far. He suggests that "we no longer know what is real or imaginary, not because they are confused, but because we do not have to know and there is no longer even a place from which to ask" (quoted in Ryan 2001 33). This stance is typical of a postmodern epistemology that has sought and fought to erase notions of difference between the virtual and the real. Whilst acknowledging that ultimately there is not a 'confusion' in the audience's mind between which is which, Deleuze nonetheless proposes that the differentiation lacks relevance. However, at least for a theatre audience, the understanding and appreciation of differentiation between the live and the virtual is key to the pleasure principle of spectatorship in digital performance. When a live onstage body encounters its projected double, the conjunction may create rich and affecting illusions, but rarely, if ever, attempts a convincing deception. This is a simple and obvious, but nonetheless important observation to make given the weight of discourses arguing to the contrary. Whilst the digital double exalts in a playfulness which conjoins the virtual and the real, the audience delights not in the indistinguishability of the two bodies, but in their own ability to understand and appreciate their interrelationship and difference (see Dixon 2003).

The digital double representing alternate embodiments or alter-egos of the performer has been an abiding source of inspiration and experimentation for The Chameleons Group, a performance company established in 1994, which I direct. In *Chameleons 4: the doors of serenity* (2002), characters undertake long and involved conversations with their projected doubles and trebles. An early sequence introduces a cyborg character, and the use of projected doubles emphasises his ontology as a schizophrenic, technologically networked and mediatised being. The live performer stands on stage watching his inverted double on screen, swaying upside-down suspended by a rope from the bough of a tree, recalling the tarot card image of 'The hanged man'. The projected scene suddenly changes to a room where the hanged man now appears in close up on a television monitor in the foreground. He speaks directly to his live alter-ego, and a short, snappy conversation ensues

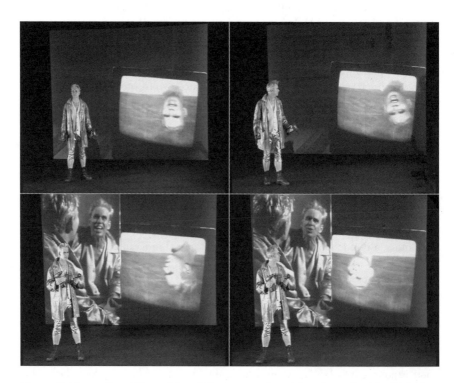

Figure 3. The author (Steve Dixon) first encounters his cyborg double ('Upside-down-boy') followed by his cyborg treble ('Mirror-boy') in The Chameleons Group's *Chameleons 4: the doors of serenity* (2002).

between stage and screen during which 'Upside-down-boy' warns the stage cyborg that he should pull himself together and forget "the metal-man bullshit".

Yet another double of the cyborg, 'Mirror-boy', then enters the pre-recorded video footage of the room to extend the play of reflections and doubles. The onstage cyborg, already doubled in his conversation with 'Upside-down-boy', is then trebled with the screen entrance of 'Mirror-boy'. In a sense, he is also quadrupled by virtue of a doubled image of Mirror-boy: his body, back towards camera within the frame, and his face and frontal reflection in the mirror. A three-way conversation ensues between the three manifestations of the same character, each of which have distinct, contrasting personalities. The dialogue is fast and comic, structured in a series of grand conceits which are then exploded bathetically by insults. It is played like a pastiche of an imagined sci-fi gangster movie:

Mirror-boy :	*So let's talk the next move.*
Upside-down-Boy :	*I want out of this crap.*
Mirror-boy :	*Oh really. Why's that?*
Upside-down-Boy :	*Well, just for starters - He's the campest cyborg I've ever seen.*
Cyborg :	*All cyborgs are camp.*
Mirror-boy :	*His metamorphosis is perfect. His advanced hyper-toned metal body cunningly disguised by his paranoid, wasted appearance and slight beer gut. Who would guess he has the strength of a hundred men and a ferrous brain capable of ten billion computations a micro-second.*
Cyborg :	*You know, I do feel particularly intelligent.*
Upside-down-Boy :	*Your brains are shrapnel, you fuckin' retard freak.*

The sequence continues in this vein, emphasising 'double' and triple perspectives on cyborgism. Mirror-boy voices utopian and messianic rhetoric, Upside-down-boy is fiercely cynical and dystopian, and the live cyborg character is an innocent and confused figure caught in the cross-fire. Another sequence features two women performing live, each with two separate pre-recorded doubles of themselves on screen, created using digital split-screen techniques. The scene is set in a public toilet and involves a complex dialogue between the six characters played by the two live and four recorded women. The sequence becomes ever more complex as the four doubles move around the space, switch places, and go in and out of the toilet cubicles, often re-emerging as the other's double, having swapped clothes.

The double as spiritual emanation

In *The golden bough*, Frazer notes the common supposition amongst primitive peoples that the soul can escape from the body through its natural openings, particularly the mouth and nostrils, and that dream is believed to be the actual wandering of the soul. Thus, "it is a common rule with primitive people not to waken a sleeper, because his soul is away and might not have time to get back; so if the man wakened without his soul, he would fall sick" (Frazer 1925 181). He relates how the story of the external soul as a concrete double is told "in various forms, by all Aryan peoples from Hindoostan to the Hebrides" (Frazer 1925 668)

Figure 4. Wendy Reed (left) and Anna Fenemore (right) hold an intense six-way conversation in The Chameleons Group's *Chameleons 4: the doors of serenity.*

and details numerous legends of warlocks and magical beings whose physical bodies remain invulnerable since their souls are hidden in secret places.

Digital doubles representing spiritual emanations or incarnations of the body relate to notions of ghosts, astral bodies, out-of-body experiences, and soul projection. The double may be depicted as a gaseous figure composed of particles, or in another, more liquid-like ethereal form, luminous and transparent. During Mesh Performance Partnerships' *Contours* (1997), Susan Kozel rolls and dances on the floor with her digital double, which is projected onto the floor from above. In the low stage lighting, the double shines out as a luminous, ghostly white figure mirroring her movements. It is a slow and sensual duet between Kozel and her double, which appears as a second, amorphous body composed of pure white light, shimmering like a spiritual aura around her. It recalls the nimbus and aureole haloes of Christian religious art, and the early photographs of Hippolyte Baraduc (1850–1909) and Louis Darget (1847–1921) showing the paranormal phenomena of etheric spirits and ectoplasm emanating from the bodies of clairvoyants.

Roy Ascott considers digital arts and telematics in relation to 'the technology of transcendence', maintaining that "the mysterium of consciousness may be the final frontier for both art and science, and perhaps where they will converge" (Ascott 1999 66).

For Ascott, digital technology also engenders 'double consciousness' and 'the double gaze'. Locating cyberspace firmly and emphatically in relation to the spiritual and liminal paradigms of Sufism and shamanism, digital artistic aspiration is conceptualised as a type of navigation through consciousness which doubles and alternates back and forth from the visionary to the material. Ascott equates this to his experiences of ingesting the sacred hallucinogenic plant ayahuasca whilst living with a Brazilian tribe, whereby he became conscious of inhabiting two bodies: his familiar phenomenological one, and a second composed of particles of light.

This image of the double body as particles is memorably played out in Company in Space's dance-theatre performance *Incarnate* (2001). A digital graphic projection of a female body shape filled with points of brightly coloured lights moves busily, expanding and contracting the size of the figure, and finally disintegrating and re-materialising the particle body like exploding nebulae. Such images emphasise digital performance's representation of the human body as a site for dynamic metamorphoses through technological intervention. The body's cycle of materialisation, dematerialisation and rematerialisation is articulated in a dynamic transformation of shimmering, star-like coloured particles. The metamorphosing flickering lights and stars composing the double in *Incarnate* represents the technologically mutating body as a cosmological and visionary form.

In Igloo's *Viking shoppers* (2000) two live dancers work in very low light upstage right, behind and to one side of a large projection screen positioned close to the audience, downstage left. A special ASCII video camera relays live images of the duet onto the screen, mapping and converting the dancers' bodies into the numbers and abstract symbols of digital code. As the dance progresses, the computer software is manipulated to decrease the density of numbers and symbols that compose the digital figures. They become less and less solid until limbs are reduced to a single thickness of code. The movement traced out becomes more and more minimal, more aesthetically hypnotic, and somehow more painful. Although the live dancers coexist on stage with the projections they generate, the digital image once again dominates: it is as though the hieroglyphic doubles are the 'real' dancers. The dancing doubles are composed of mesmeric shimmering symbols, like a high-tech encapsulation of Artaud's proscriptions for a theatre alive with 'physical hieroglyphs'. There is a sparse but intensely beautiful simplicity to these doubles, which seem at once material and ephemeral. The performance is Brechtian in revealing its computational mechanics of production, but at the same time, this 'making strange' is deeply mystical: the doubles are like etheric phantoms composed of particles of cybernetic dust.

Troika Ranch's *The chemical wedding of Christian Rosenkreutz* (2001) exemplifies the digital double as a spirit form. It explores the mystical union of computers and human beings, taking its inspiration from a 17th century spiritual allegory tracing Rosenkreutz's transcendent transformation through the 'chemical wedding' of alchemy. Troika Ranch have described how the sensory technologies they employ are "the 'magic' of our time" and how the audience should perceive this aspect of the performance as magic—just as alchemy was viewed in the 15th century. The company combine dance and dramatic dialogue, and use their custom-built motion sensing system *MidiDancer* (see this volume pp. 5–12). This uses flex sensors attached to the dancer's joints (elbows, knees, wrists, hips etc.) which are attached by wires to a single-chip microcomputer in a small box usually worn on the dancer's back. The computer measures the degree of 'flexion' of each

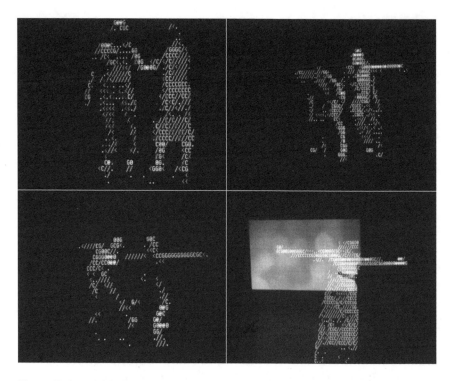

Figure 5. An ASCII camera converts the body shapes of two live dancers into computer code in Igloo's *Viking Shoppers* (2000).

joint thirty times per second, assigning it a number between zero (straight) and 100 (fully bent), and transmits the information via radio waves to computers off stage. These 'intelligently interpret' the data and trigger technical cues and manipulate audio and visual media from hard disc files, or from the input of a live video camera. These include video projections of the performers' digital doubles, which are mutated to render them luminous silhouettes, to split them into consecutive frames, and to shatter them into shards.

The piece is spiritual in tone and choreographic style, which makes use of numerous dance 'lifts' and other symbolic 'ascents', as well as ritualised, secretive gestures. Arm and body movements frequently come to rest in poses redolent of ancient religious hieroglyphics, once again recalling Artaud's calls for a 'doubled' theatre of 'physical hieroglyphs'. The projected imagery fuses surrealism with mysticism. As a performer dances solo on stage, a slowly moving composite image of twenty staring eyes is projected, superimposed with a live video feed of the dancer's double, which repeats her lyrical, graceful movements with a one second delay. In another sequence, the performers' doubles appear one by one on screen, floating serenely down through a black space, their naked bodies slowly rotating. Sometimes one large single figure, sometimes five or six in varied perspective scales as though near and far, these spiritualised, ethereal bodies descend majestically through space at different speeds. Many of the figures' arms are cast heavenwards, in classical, celestial pose.

The digital double as a soul or spiritual emanation represents not the split subjectivity of postmodern consciousness, but rather a symbol of the unified, cosmic and transcendental self. Discourses on cyberculture and digital performance have long been obsessed by the Cartesian split between body and mind, and the polarities of absence and presence, real and virtual. The performance of the digital double as expressed by groups such as Mesh Performance Partnerships, Company in Space, Igloo and Troika Ranch, offers a different and in many ways more fundamental binary for digital performance studies: between the corporeal and the spiritual.

The double as manipulable mannequin

The computer avatar, a graphical 'stand-in' for the human body within virtual worlds, links the notion of the double as a spiritual emanation to the final category of digital double, the manipulable mannequin. The term 'avatar' derives from Hindu scriptures, being the bodily incarnation of deities. The Sanskrit *Avatara* translates as a descent, the passing down of the gods from heaven to the material world, and artists such as Mika Tuomola have created avatars which embody this sense of a mythical being or digital deity. Tuomola (1999) has consulted on and developed numerous imaginative and dramatic avatar worlds, drawing on ideas from dream, myth, carnival and Commedia dell'Arte.

However, as Gregory Little (1998) observes, the vast majority of avatars inhabiting online environments are not personal representations of the user, but clip-art imports from hegemonic multinational portfolios: Disney characters, Barbie dolls and Baywatch babes, pop icons, Stallone war-machines, smiley faces, and consumer objects like Nike trainers. Lex Lonehead points out that even when users design their own avatars, they are "mostly supermodel cutouts and cutesy cartoons" (quoted in Little 1998). Little argues that "the mind/body dichotomy is a red herring" and that the most dangerous binary is the pairing of self and commodity. In Little's analysis, people's identification with avatars constructed from trademark iconography and other commodified imagery thus represents our ultimate desire to become commodities. Rather than creating new virtual bodies that symbolise our mythic fantasies or personal desires, avatars "function as a top-down tool, the embodiment of post-modern, multinational commerce.... It is Kapital, not Krishna that makes itself a body".

In his *A manifesto for avatars*, Little urges the creation of radical new avatars which resist integration and stability, explore taboos, counter myths of transcendentalism and resist the internal/external binary. These are Artaudian manifestations: "images that speak of hyperembodiment, of extremes of physicality, like the visceral, the abject, the defiled, and the horrific". Little discusses Artaud's experience of losing conscious control of his body, and his description of the experience as 'the body without organs'. Artaud's concept pre-figured Deleuze and Guattari's much-cited use of the formulation in *Anti-Oedipus* and *Mille plateaus*, where they characterise it as unproductive, sterile, unengendered, and unconsumable. Little suggests that the body without organs, just like the avatar, "mirrors post-biological structures that undermine anatomical classification, capitalist consumption, and tedious mind/body/commodity separations in support of a more distributed, nomadic, and emergent model of embodied consciousness".

The double as a computer-generated 'body without organs' features strongly in dance performance, where software applications such as *LifeForms* are now commonly used to manipulate graphical figures on the desktop to conceive and experiment with choreography prior to studio work with live dancers. Motion capture technology similarly renders a manipulable double of the dancer onto the computer screen. For sequences in Merce Cunningham's *Biped* (2000) dancers worked in a studio surrounded by video cameras, which tracked the movements of reflective balls attached to their body suits. The computer artist, Paul Kaiser, used software to 'join up the dots' between the nodes and build up a solid 'biped' virtual body shape around them. The data was then manipulated into an animation of bold, coloured lines, giving the impression of hand-drawn dancing figures, composed with shape and line, but without solidity. During performances of *Biped* these ghostly doubles were projected onto the transparent stage scrim as giant 30-foot figures, to interact with the live dancers.

Yacov Sharir combines *LifeForms* and *Poser* software to choreograph beautiful virtual dancers which are created from start to finish within the computer itself, without any digital input taken from human dancers. He calls his screen figures 'cyber-human dancers', which defy gravity to float, pivot and fly through dramatically coloured and rendered three-dimensional virtual spaces. Live performers have also used computer technologies to convert their own physical bodies to the condition of a manipulable mannequin. In Marcel-Lí.Anthúnez Roca's solo performance *Epizoo* (1994) audience members use touch-screen computers to activate robotic manipulations of his body. Pneumatic metal moulds and hook devices are attached to his head and body, which pump up and down against his buttocks and chest, wiggle his ears, and pull his mouth and nostrils open and shut. The performance presents an at once comic and savage representation of the helpless, dehumanised, mannequin body at the mercy of technology, and comments on the ability of humans to use technology to electronically torture others at a distance.

But if there is one single, powerful image that will remain within the history of early digital performance, I believe it is of Stelarc, his naked body jerking spasmodically and involuntarily in response to electrical impulses sent along the Internet. In a number of different performances during the 1990s Stelarc wired himself up to a computer to become a manipulable mannequin, controlled by audience members and Internet traffic. Through different interface systems, signals sent via the Internet remotely stimulated muscles in different parts of his body via electrical sensors, activating a startling and macabre physical performance. For *Fractal flesh* (1995), audiences in Paris, Helsinki and Amsterdam simultaneously used graphical images of his double on touch-screen computers to activate different areas of his body, and in *Ping body* (1996) the live flow of Internet activity was used as stimulation. During the *ParaSite* (1997) performances a customised search engine scanned the Internet retrieving medical and anatomical pictures of the body, then mapped the jpeg images onto his muscles to induce involuntary motions. As the performance took place Stelarc's movements also fed into a VRML (Virtual Reality Modelling Language) space at the venue, and onto a website. In *ParaSite*, says Stelarc:

The cyborged body enters a symbiotic/parasitic relationship with information … the body becomes a reactive node in an extended virtual nervous system … the body, consuming and consumed by the information stream, becomes enmeshed within an extended symbolic and cyborg system mapped and moved by its search prosthetics.
(Stelarc 1999)

Stelarc's (in)famous declaration that 'the body is obsolete' is made in the light of what he considers to be far superior technologies which have developed in different forms and hardware embodiments. Stelarc's recent cyborgic performances such as *Exoskeleton* (1999) where he rides aboard a six-legged pneumatically powered robot, celebrate the conjunction of human flesh with 'intelligent' metal. Other artists now omit the flesh part of the equation to leave the performing entirely to humanoid robots, which constitute the most advanced manifestations of the human double. Anthropomorphic robots actuate Edward Gordon Craig's (1911) early twentieth century call for the replacement of the actor by the Über-marionette and, as Margaret Morse (1998) observes, they often bear remarkable resemblance to their creators. The notion of robots as doubles or clones of their artist inventors is also stressed in Jeff Cook's discourse on robots in relation to male procreation, constituting a "patriarchal dream of parthenogenesis" which escapes "the indeterminate and unreasonable realm" of flesh, nature, and the feminine (Cook 1997).

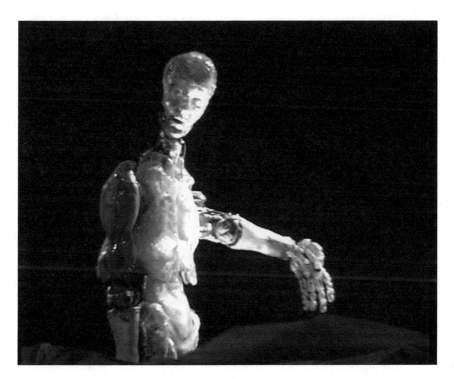

Figure 6. One of some sixty robot 'performers' in Amorphic Robot Works' *The ancestral path through the amorphic landscape* (2000), designed and directed by Chico McMurtrie.

From Nam June Paik and Shuya Abe's early female performing robot *Robot K-456* (1964) to the recent advanced creations of robotic artists such as Chico McMurtrie and Mark Pauline, these figures constitute the double as a manipulable mannequin in its most distinctly high-tech and solid form. Whilst the robots' manipulation as a mannequin is primarily defined by the artists' programming of its repertoire of actions, interactive devices have also enabled audiences to engage with the robots and introduce their own manipulations. Norman White's *Helpless robot*, or *HLR* (1988) is a dual-computer controlled talking robot which has been exhibited in numerous venues including shopping malls. Infrared motion sensors detect people nearby, and *HLR* begins to speak to them, asking them to physically help to move it around its metal axis base. As the person engages with the robot and begins to manipulate it, *HLR* becomes increasingly annoyed, accusing the person of putting it in the wrong position, and telling them to move it further left, then right, then left, then right. Whatever the spectator does, it continues to abuse them, soon yelling and berating them for hurting it. The more the helpful spectator attempts to correct *HLR*'s position, the louder the abuse and the fiercer the accusations. When the spectator finally gives up and moves away, *HLR*'s tone changes into a pathetic whine. "I'm sorry. Please come back. I'll behave." it says, and the cycle begins again. Arthur Kroker offers a humorous critique of *HLR*, relating it to Baudrillard's technological theories. Both the robot and Baudrillard's texts at first "sit quietly in the midst of the frenzy of the mediascape" but then "invade your mind like a media virus". Liberation from both entails "brain vacuuming: the mental shutdown which Nietzsche once described as the dominant activity of the 'last man'" (Kroker 1992 65).

Artaud's calls for the use of giant stage mannequins in *The theatre and its double* have thus been answered in myriad ways within digital performance, from avatars and robots to virtual dancers, to the reduction of the live human body itself to a puppet, manipulated by audiences at a distance. The manipulable mannequin acts as a design tool for conceiving performance in virtual form prior to work with human performers, as an avatar or Über-marionette replacement for the live performer, or as the performer herself, at the mercy of digital manipulations.

Conclusion

The digital double is a mysterious figure which takes various forms and undertakes different functions within digital performance. The reflection double announces the emergence of the self-reflexive, technologised self, conceived as becoming increasingly indistinguishable from its human counterpart. The alter-ego double is the dark doppelgänger representing the Id, split consciousness, and the schizophrenic self. The double as a spiritual emanation symbolises a mystical conception of the virtual body, performing a projection of the transcendent self or the soul. The manipulable mannequin, the most common of all computer doubles, plays myriad dramatic roles: as a conceptual template, as a replacement body, and as the body of a synthetic being.

The notion of the 'other' was first discussed at length by Freud, and later became embraced by social theorists to denote various understandings of cultural difference following Edward Said's reorientation of the word in *Orientalism* (1978).

In his article *Screen test of the double*, Matthew Causey discusses the moment when the double appears and the live actor confronts his or her digital 'other'. Drawing on Freud's notions of the uncanny and Lacan's theory of the mirror stage, he suggests that this representation of the self outside itself, the witnessing of the self as other:

> constitutes the staging of the privileged object of the split subject ... enacting the subject's annihilation, its nothingness. ... The ego does not believe in the possibility of its own death. The unconscious thinks it is immortal. The uncanny experience of the double is Death made material. Unavoidable. Present. Screened".
>
> *(Causey 1999 385–86)*

Causey's understanding of the power of the performer's double as a multimedia image is revealed, first and foremost, as a *memento mori*. In a later article, he returns to the theme again to present a chilling image of the performer's double as cadaver: "to see one's self is to demolish oneself in an autopsy of perception" (Causey 2002 65). Drawing on *Cinema I* where Deleuze's discussion of the three varieties of film's 'movement-image' ends with "death, immobility, blackness" (Deleuze 1986 68), Causey conceptualizes the performer's confrontation with their double as "the quest for disappearance ... a quest for otherness" (Causey 2002 66).

Causey's reading of the modern, high-tech double as a symbol of death can be related back both to the Narcissus myth at the centre of my first example, *Liquid Views* (Narcissus languished and died through his self-hypnotic gaze on his reflection) and to ancient and primitive beliefs about the double. In *The golden bough*, Frazer recounts numerous examples of cultural fears associated with seeing one's reflection in water. In ancient India and Greece it was a maxim never to look into still water, and to dream of seeing one's reflection was considered an omen of death. The Zulus dare not look into a dark pool for fear that a malevolent water spirit may capture the reflection, and the Basutos believe that a crocodile has the power to kill a man by dragging his reflection under water (Frazer 1925 192).

These ancient beliefs cast a sinister shadow over the digital double and our fascination with mediatised reflections. Whilst Artaud pronounces the double as theatre's spiritual becoming and writers such as Ascott conceive the technological double in terms of transcendental metaphysics, others point to its dark menace and terminal nature. Scientists such as Kevin Warwick (1997), Ollivier Dyens (2001) and Rodney A. Brooks (2002) predict that technological progress will culminate in the ultimate triumph of the double, in the replacement of humans by intelligent machines. If this proves to be the case, the beautiful and psychologically compelling depictions of the human double we see within digital performance may not simply be reflections, alter-egos, spiritual emanations and mannequins, but dark sirens leading us seductively to the same mesmerised paralysis and ultimate death that was Narcissus' fate.

References

Artaud, A. ([1938 in French] 1974) The theatre and its double. In Artaud, A. *Collected works, Vol. 4*. Calder Publications, London.

Ascott, R. (1999) Seeing double: art and the technology of transcendence. In Ascott, R. (ed.) *Reframing consciousness: art, mind and technology*. Intellect, Exeter, pp. 66–71.

Brooks, R. A. (2002) *Robot: the future of flesh and machines.* Penguin, London.

Bell, P. (2000) Dialogic media production and inter-media exchange. *Journal of Dramatic Theory and Criticism,* Spring 2000, 41–55.

Causey, M. (2002) The aesthetics of disappearance and the politics of visibility in the performance of technology. *Gramma* **10** 59–72.

Causey, M. (1999) Screen test of the double: the uncanny performer in the space of technology. *Theatre Journal* **51**(4) 383–394.

Cook, J. J. (1997) *Myths, robots and procreation.* http://murlin.va.com.au/metabody/text/mythrobopro.htm. Accessed June 2001.

Craig, E. G. (1911) The actor and the über-marionette. In Craig, E.G. (1911) *On the art of the theatre.* Theatre Arts Books, New York, pp. 80–94.

Deleuze, G. (1986) *Cinema I: the movement-image.* Trans. Tomlinson, H. and Habberjam, B. University of Minnesota Press, Minneapolis.

Deleuze, G. and Guattari, F. (1987) *A thousand plateaus: capitalism and schizophrenia.* Trans. Massumi, B. University of Minnesota Press, Minneapolis.

Derrida, Jacques ([1970] 1981) *The double session: dissemination.* Trans. Johnson, B. Chicago University Press, Chicago.

Dixon, S. (2003) Absent fiends: Internet theatre, posthuman bodies and the interactive void. Performance Arts International, *Presence* online issue. http://www.mdx.ac.uk/www/epai/presencesite/index.html. Accessed July 2003.

Dyens, O. (2001) *Metal and flesh: the evolution of man: technology takes over.* MIT Press, Cambridge, Massachusetts.

Frazer, Sir J. ([1922] 1925) *The golden bough: a study in magic and religion.* Macmillan, London.

Freud, S. (1985) The uncanny. Trans. Strachey, J. In Freud, S. *Pelican Freud library, Vol. 14.* Penguin, Harmondsworth, pp. 339–76.

Heidegger, M. (1962) *Being and time.* Trans. Macquarrie, J. and Robinson, E. Blackwell, Oxford.

Herzogenrath, B. (2002) Join the United Mutations: Tod Browning's Freaks. *Post Script* **21**(3) 8–13. Also available at http://courses.lib.odu.edu/engl/dpagano/freaks1.doc. Accessed June 2003.

Hoffmann, E. (1982) The sandman. In Hoffmann, E. *Tales of Hoffman.* Trans. Hollingdale, R.J. Penguin, London, pp. 85–126.

Holzer, H. (1974) *Encyclopedia of witchcraft and demonology.* Octopus Books, London.

Kroker, A. (1992) *The possessed individual: technology and the French postmodern.* St. Martin's Press, New York.

Lacan, J. (1977) The mirror stage as formative of the function of the I. In *Ecrits: a selection.* Trans. Sheridan, A.. Norton, New York, pp. 1–7.

Little, G. (1998) *A manifesto for avatars.* http://art.bgsu.edu/~glittle/ava_text_1.html. Accessed November 2001.

Morse, M. (1998) *Virtualities: television, media art and cyberculture.* Indiana University Press, Bloomington and Indianapolis.

Royle, N. (2003) *The uncanny.* Routledge, New York.

Ryan, D. (2001) Painting and its double: David Reed and Franz Ackermann. *Contemporary Visual Arts* **34** 23–26.

Said, E. W. ([1978] 1995) *Orientalism: Western conceptions of the Orient.* Penguin, London.

Stelarc (1999) *ParaSite: event for invaded and involuntary body.* http://www.stelarc.va.com.au/parasite/index.htm. Accessed August 1999.

Tuomola, M. (1999) Drama and the digital domain: Commedia dell'arte, characterisation, collaboration and computers. *Digital Creativity* **10**(3) 167–179.

Warwick, K. (1997) *March of the machines: why the new race of robots will rule the world…* Century Books, London.

4 The emergence of hyper-reality in performance

Kjell Yngve Petersen

The emergence of hyper-reality in performance

This chapter describes how I think performance informed by real-time and telematic technology promotes the emergence of a new hyper-reality stage language. I am limiting my enquiry to the visual aspects of the performance experience and thus limiting the complexity of my arguments. I will, through my writing, introduce a dynamic semiotic view on performance and thus redefine or reformulate the basic phenomena related to performance and the experience of 'the human real'.

I have a special interest in how the use of real-time and telematic technology in performance makes it possible to develop a poly-focal approach to the staging of performances, involving many times, places and dimensions at the same time. And how a new reality construct evolves from the stage montage of these not originally related and synchronous events, combining asynchronous and parallel occurrences into a 'hyper-reality'. In the frame of this new 'reality' new paradigms of the stage language can be established; new paradigms for what phenomenon and themes can be represented and discussed on stage.

'Moving imprints of action': the visual aspect of 'telematic performance'

My definition of a 'telematic performance' uses a multi-media and data memory space where the live performance contributes through cameras and sensors with media and data, which then individually is combined and processed into 'views on reality' which are then represented in the stage environment as a montage of views of realities. The very definition of 'telematic' is the combination of tele-action, tele-presence or communication technology with the computer, that enables an asynchronous relationship between spatially or otherwise distanced locations or events.

I will focus on the use of cameras, computers and video-projectors in the stage setting, and thus discuss the visual aspects of performance being infused by telematic and real-time technology. This digital visual technology has developed from the cinematograph and can fruitfully be thought of as 'moving imprints of actions'.

Cinematograph was one of the original names for 'moving pictures' appearing at the end of the 19th century. It is a compound Greek word:

Kinemâ = *movement; movements of bodies without consideration to what is moving nor to the cause of movement; light and the motion arts for example illuminating moving sculptures.*

The second part of the word comes from:

–graf - *writing, a means of producing records, signs and imprints.*

Its combined meaning is: a machine that shows 'moving imprints of actions'.

The performers are recorded visually—the material is processed and used to create pictorial space on stage. In doing so the performers leave behind actions as data memory—and copies, duplicates or interpretations are present as projections into the scenography. These visualisations are actions, which have left the body and now exist on their own terms as projections. In this way we consider the filmed action to be something that detaches itself from its origins and become an independent entity. As this situation develops it points to a dilemma between the motivation of the action and its interpretation, that is between the intention of the action being carried out by someone, and the montage of the interpretation of the visual imprints processed by art machines and projected back into the stage setting. Through this a new form of stage reality occurs when several individually consistent but not directly related representations of 'reality' are experienced simultaneously. Our normal cognitive-semiotic construction of a reality is performed by combining a complex of experiences through interpretation into a meaningful whole. In telematic constructions new possibilities of sensing and acting get involved, and new mixed forms of what reality can be, are established. The notion of reality can in these new settings be a montage of several times, places, viewpoints and interpretations within the same stage reality.

The advent of real-time and telematic technologies in the stage space decisively changes the fundamental conditions of performance, while making it necessary to reconsider many of the basic concepts used in understanding the theatrical situation. These techniques suspend the concepts of space and time as fixed definitions, thus making it possible to manipulate the spatial and temporal parameters of the performance. Instead, a kind of fluid reality emerges, in which space can be time-like and time can be space-like; a kind of augmented space-time where displaced or parallel instances of time and space are possible. Even some of the basic concepts of performance lose their clear and distinct meanings, such as it being a 'live' event and that everything is realised for 'real' in real-time and real-space with real consequences.

Memorandum. Dumb-Type, Japan, 2000

An example of the montage of a time perspective is a scene in the play *Memorandum* (2000) by the Japanese performance group *Dumb-Type* (http://dt.ntticc.or.jp). On the stage a 5-minute scene is played in a setting with a few pieces of furniture. On a broad back wall we see four projections showing four

Figure 1. HyperReality. Images of *Memorandum*. Performed in Japan 2002. ©DumbType 2002. Photographers: (top) Kazuo Fukunaga, (bottom Yoko Takatani.

reflected versions of the action on the stage in almost full size. The four projected versions of the stage-action replays the exact same sequence, but at four different speeds—one very slow, one slow, one fast and one very fast. They have been edited to reach the climax of the action simultaneously at the end of the scene. As audience we in this way experience five different versions of the same act. Time has been stretched out and established as five parallel spatialities all going through the same actions and reaching for the same climax but at different pace. In this scene time becomes space-like and the stage setting becomes a montage of several simultaneous time-views spread out in space.

This is neither real-time or telematic in its technical construction, but it is using our knowledge as audience of real-time and telematic phenomenon to stage in the theatrical universe of performance an expression through telematic occurrences.

The evolution of realities in performance into hyper-realities

Through the advent of real-time and telematic technologies in the stage space (Ascott 2003) it becomes possible to develop polyfocal viewpoints and scenic montages with multiple perspectives, which in turn makes it possible to work with an approach to the staging of performances which involves many times, many places and many dimensions in one and the same narrative. When this is done in computer-controlled real-time and telematic processes, the stage becomes a self-reflecting, generative system. The stage becomes 'hyper-mediumised', and the stage language becomes a hyper-montage of differing levels of abstraction of itself. The stage reality is then constantly commenting and redefining its own notion of reality, and becomes a hyper-reality situation.

In this situation, it becomes of interest to consider the performance as a semiotic realism—as dynamic meaning-forming processes, which organise our perception of the real—what we can call the construction of the 'human real'. This is a notion coined by Professor Per Aage Brandt as a part of his research into a non-reductive, semiotic realism in the field of dynamic semiotics. To begin with we will look at the performance situation, as it can be understood from the field of dynamic semiotics.

Performance as a formal sign system

One of the fundamental notions in performance, on which all strategies and methods are based, is that people who are actually present execute it. It is a question of an artistic event, which only exists and takes place through the actions of the artists themselves. These actions are bound in time and space and are executed live in real-time and real-space. The point of departure is the simple observation that, as human beings, we are our bodies. It is within the body that we exist, and it is from the body that our existence unfolds. It is this authenticity—the fact that some one is actually doing that which we (the audience) perceive—that separates performance from all other art forms. In performance, the experience is given immediacy by the performer and every single event is unassailably real (Barba 1994).

Although the performance is based on the presence of the body, it is also a form of play with the body removed into the presentation from the physical binding—the formal use of the body as the bearer of signs. The body is used to communicate through 'linguistic motorics' and the 'performative discussion' is used to expand our perception of reality and our mental and physical relationships to it. This creates a sense of mutuality and human relationship, the focus of which is the theatrical and in which meta-fictional strategies are the means of communication.

The introduction of real-time and telematic technologies into the performance situation results in major changes to the tools available for articulation, and rearranges and increases the modes of abstraction and complexity in the stage language. This moves the focus of the creation strategies and the understanding of what can be formulated in performance from a theatricalisation on notions of ordinary reality, to a reconstruction of the 'performative language' on a more basic semiotic level primarily dependent on the way in which we, as people, construct 'the human real'.

Performance operates by constructing realities. Performance can be considered to consist of special ritual situations, in which the actions are part of a meta-fictional linguistic universe and in which, therefore, the performers should not be understood to be executing purposeful actions. The art of performance can be understood as the formal stratagems which work with the human real and which operate in the actual processes that shape the comprehensible.

The construction of 'the human real'

What are the relationships between thought, speech and action in our construction of 'the human real'—and how can we understand the concepts of theatricality and meta-fiction as dynamic, semiotic phenomena?

We receive the world as chaos, a vast stream of events and impressions. We experience, sort and analyse them, so that they become a comprehensible, ordered past, as traces of validated memory. Reality then emerges when the traces of the past meets the flow of the present in the action of expectancy (Brandt 1995). This then happens in a gradually increasing process of abstractions from physical to meta-fictional levels.

The notion of a relationship to reality can be said to arise when we hesitate, when we develop a space of time between action and inaction, in which degrees of meta-action can emerge and become physical linguistics, for instance, gestural meanings. If we hesitate a little longer and utilise the motoric capacity of our speech organs, vocal actions develop into speech. We speak and intone because and to the extent that we hesitate. The development of body language and spoken language is a question of how much we hesitate. Priorities and strategies of the authority of meanings are inherent in body language's use of gesticulation and the spoken language's use of vocalisation, and both of the latter are capable of developing into articulated linguistic tools. We can then practise mutual hesitation, i.e. we converse (Brandt 2002). If we hesitate for an extended period, we 'think'. We see that others are thinking because we see them hesitate.

By thinking and communicating, we learn to adapt ourselves to each other's concepts and a new form of reality emerges; that of meaning. We start to play. When we hesitate and play, we establish something that is more than mere presence and action; we attain meanings and form opinions. We play reality into existence.

Theatricality emerges when play develops into conscious performance. Theatricality develops from the fundamental processes in which man considers his actions. This is the testing ground where possible events are acted out in a ritualised framework, so that we can weigh their consequences without our actions becoming reality or having real consequences. We play and simulate. The actions are executed relieved from their purpose and transposed into different levels of abstraction. The transitions between action and simulated action are where we test, consider and experience the potential of our actions. This is where the possibility of the intentional action emerges—of action that tests the potential for the comprehension of action in the relationship between experience and action. This is the intermediate performative space where the relationship in the interstices between theatricality and reality is scanned and developed into processes and structures. It is the process of the playful human playing with his playfulness, and thereby developing and testing abstract notions of its relationship to the world and itself.

The reality of the performance is fiction. From this basic starting point the language of the performance is used to generate a meta-fictional relationship to the stage as the reality of the fiction. This development is pursued by emphasising the stage's true reality through a game, which points to the fiction as reality. Non-theatrical acts are carried out within the theatrical situation, acts which refer back to reality as a concept. Performance can in this way be said to be a formal treatment of the relationship between theatricality and reality. Performance is a phenomenon in which we can examine and develop on our construction of 'the human real'.

Polyfocal views and multiple perspectives

Such tools as cameras and projectors on the stage make it possible to collect impressions and deliver expressions. How, then, is the performer perceived in this landscape of real and virtual presences?

It is possible to set up actual reflections in which the performer is reproduced directly and life-sized. In this case, video projection will be perceived as a pure, semiotic reflection (Eco 1986), i.e. that something actually occurring in real time is being reflected. We are familiar with the effects of mirrors and accept the projection as a pure, semiotic channel, that enables us to see from a different viewpoint. The projected image of the performer can then be considered real—we simply monitor his actions through the camera and projection. So far we are still in a coherent universe where the previous notions of the language and reality in performance is preserved.

But we cannot know whether it is an earlier recording or whether the medium stream has been processed along the way. However, since everything in a performative situation exists in varying degrees of abstraction, we can manipulate time, location, image segment and camera motion to a quite considerable extent, and the projections will still be perceived as expansions of the real, rather than as fictional presentations. This is further underpinned by the fact that our experience of the mass media and surveillance systems has given us a calibrated approach to the state of images, i.e. when something is genuine documentation, when it has been manipulated and when it is fiction. Thus, as an observer it becomes interesting to decode the degrees of reality and the phenomenal relationships between different representations. We can ask ourselves whether the performer is a living person—who is present and who seeks to become abstract through his theatrical gestures—or whether the projected figure is a media representation—who seeks to become real through reference to the filmed person.

The breaking up of the stage's logo of place is further expanded when we introduce additional cameras and start to form and combine the images during the recordings. The potential arises for seeing a performance from several angles and in several sections at one and the same time. We can work with polyfocal views on the stage, and they can be arranged into montages of poly-perspectives.

By inserting computers into the media stream, making it a telematic construct, we can create displacements in time—capture an action and display it later, display it more slowly, manipulate the image content, etc. We attain what can be called a multiplicity of temporal foci, locational foci and interpretational foci. A montage in multi-perspective can therefore be a simultaneous presentation of differing interpretations of one and the same motif, thereby establishing a multi-interpretational perspective.

The performative work's composition becomes a space of potentiality, of several simultaneous viewpoints with several simultaneous interpretations. A dynamic interweaving of previously separated phenomenal worlds takes place, in which the individual phenomena can be freely interchanged and combined. The performative situation as a whole, cohesive event with a 'conceptually central perspective' becomes disturbed. It becomes a situation with many possible perspectives, viewpoints and reflections—a situation that promotes relational comprehension in a selective construction of the experience.

The observer's role is changed, from 'being the perceiver of' and 'identifying with', to that of 'being a co-creator' at the level of the very formation of meanings.

All parts are mutually susceptible and they form the springboard and framework of each other's development. The performance becomes a system, which, physically and perceptually, repeatedly interprets and remodels the foundations of its own interpretation. The performance setting becomes a self-reflecting, generative system, and the performance language becomes a hyper-montage of differing levels of abstraction of itself. The 'performance reality' is then constantly commenting and redefining its own notion of reality, and becomes a hyper-reality situation (Qvortrup 2003).

Many of the linguistic motorics formerly being a part of the domain of the performer is challenged and evolved by the real-time and telematic technologies. The delicate notions that can be executed by the live performer through equilibristic control of the articulation of hesitation, and the performer's elaboration on the aspects of physical space and time are transformed into being part of a relational linguistic between a physical and a mediated worldview. The form-related potential of the use of many viewpoints and montages of many perspectives are dramatically developed when used in a performance in combination with telematic tools and real-time digital processing of the medium. Then the performance becomes an evolutionary, organic, multi-dimensional language of form.

Kinematograf. Boxiganga, Denmark, 1995

My first attempt to investigate the possibilities of combining physical and media realities was the performance *Kinematograf* from 1995 (©boxiganga www.boxiganga.dk).

As mentioned, the word *Kinematograf* is derived from the Greek and can be translated as 'moving imprints of actions'. And that is what this work is about: the relationship between a physically present performer and her interaction with imprints of herself in a virtual world.

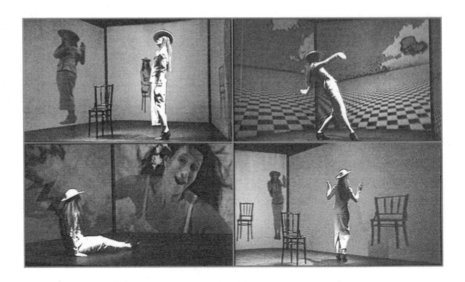

Figure 2. HyperReality. Images of *Kinematograf*. Performed in Aarhus, Denmark 1995. ©boxiganga 1995.

The scenography is designed as a classical proscenium, but with a highly exaggerated perspective. The angles of the ceiling, floor and walls are exaggerated and give the illusion of looking into a three-dimensional virtual world. Films are projected onto the three sides of the stage setting and are also projected onto smaller objects in the stage space. A single live performer, who confronts a large number of copies of herself in the virtual world, performs the work.

This work focuses on the dichotomy between the physical and the fictive, and a number of situations are enacted around this theme throughout the performance.

The performer wanders between the real and virtual worlds. She leaves copies of herself, dances with her mirror image and plays together with a series of her doubles. She carries herself around as a miniature and is seduced by demon-like replicas of herself.

It is an investigation through a formal approach to the boundary between reality and illusion, an investigation of the fictive and the real and of the relationship between authenticity and artificiality. Each of the two worlds mutually interprets the presence of the other—in relation to that which is phenomenally different and to the content of the action.

It is a performance about a person engaged in a struggle to define her reality, to cope with the fictive and the real.

This performance was produced with non-real-time technology as a fixed play between the performer and pre-recorded and edited films. On this level it was possible to stage a discussion of the linguistic possibilities of real-time and telematic technologies, but it was not possible to establish a stage reality that was actually functioning and being perceived as real-time and telematic.

Figure 3. HyperReality. Images of *TheatreResearch*. The National Danish Theatre School 2002. ©boxiganga 2002.

Labyrinth. Research at The Danish National Theatre School, 2002

This became possible when, in 2002, we started research in hyper-reality performance at The Danish National Theatre School.

In the following example—a test performance in the theatre-laboratory—an on-stage camera records the stage space from the side and the image is projected back onto the stage setting. The stage can thus be viewed from two sides at once.

The setting uses a large screen at stage centre, where the life-sized, projected performers appear. There is access to the back of the screen from both sides. A camera located to the right of the screen records the events in progress in front of and behind the screen seen from the side. It is this recording that we see projected onto the screen.

The video signal from the camera is processed by a computer, in which software makes real-time image manipulation possible. The computer enables a memory space, a space of time, where recordings of the live performance are processed and combined into views on the stage reality.

Based on simple recording and replaying, thus enabling time displacement, combined with real-time playback as a direct reflection, the performance is expanded by a large set of stratagems which constantly displace and mix views on reality. There are temporal shifts, events are played backwards and the performers mime to the projections. It is a labyrinth both as a physical space and as a mediumised displacement of time, duration and place. It is a hyper-reality situation where the stage reality continuously reforms and reinterprets itself through a hyper-montage establishing an augmented space-time.

In this setting the structure and components of the stage language is moved—from the performers' physical interpretation of their relationship with the physical space and linear time, toward a complex relationship between several equally present linguistic articulations on the stage reality. The space of hesitation—as the articulated stretch in space and time giving expressive fullness in the relationship to reality—is now also being a part of the relationships in the hyper-montage of the performance, re-configuring and expanding the linguistics of the language of the performance. When concrete space and time is no longer an untouchable basis on which the performance language is build, then other more abstract or functional manifestations can become the new basics. This can then be redefined with each new performance—and even be altered within a performance as part of the performance narrative.

References

Ascott, R. (2003) *Telematic embrace*. University of California Press, London.

Barba, E. (1994) *The paper canoe: a treatise on theatre anthropology*. Routledge, London.

Brandt, P. A. (1995) *Morphologies of meaning*. Aarhus University Press, Denmark.

Brandt, P. A. (2002) *Det menneskeligt virkelige (The human real)*. Rævens Sorte Bibliotek nr: 55, Politisk Revy, Denmark.

Eco, U. (1986) Semiotics and the philosophy of panguage. Advances in semiotics. Indiana University Press, Bloomington, pp. 202–226.

Qvortrup, L. (2003) *The hypercomplex society: digital formations*. Vol.5. Peter Lang Pub., New York.

5 The technology of the real: Wayne McGregor's *Nemesis* and the ecology of perception

Jackie Smart

The technology of the real

Wayne McGregor has said that his choreography for Random Dance "places concepts of the body, time and space into fresh dimensions" (McGregor 2000). These three concepts, in themselves and in their relation to one another, frame our human experience of the world, constructing and defining our understanding of what constitutes reality. The notion of a theatrical or performative reality which differs from that of our everyday reality is nothing new. Indeed, as Richard Schechner[1] and others have shown, audience acceptance of the 'symbolic' dimensions of theatrical time, space and body is central to the discourse of performance. We understand that, within theatre, time-space may be compressed, expanded or altered to enable entry into incorporeal dimensions such as the emotional, the psychological or the fantastic. Traditionally however, the reference point which enables us to recognise 'where we are' has usually remained human experience, by which I mean that which is rooted in body or mind. Postmodern performance takes us one step further, referring us not to an 'original' experience but to a simulacrum, an already-mediated version of reality. Wayne McGregor's particular interest lies in the technological universe and the questions he poses concern the dimensions of that universe in their relation to the human 'real'. I examine here how his choreographic explorations of technology attempt to provoke new perceptual pathways, challenging us to reconfigure our comprehension of the body, time and space by extending the location of these concepts into 'extra-human' areas. I focus on Random's production, *Nemesis*, which is their first in their new position as company in residence at London's Sadler's Wells Theatre and also celebrates a decade of work. Rather than using this event in a retrospective fashion as an opportunity to consolidate and define, McGregor conceives of *Nemesis* as a 'provocation' and of the residency overall as an opportunity to "raise new choreographic questions for ourselves and our audience" (Random 2002 video).

It is interesting that the interrogation McGregor undertakes in *Nemesis* arises to some degree from a return to basics. Previous pieces such as *Sulphur 16* and *Aeon* locate the audience from the start within complex 'virtual' spaces: *Aeon* begins in darkness with the projection onto three separate screens of abstract, white shapes which are clearly computer-generated; because of the positioning of the screens, these obscure shapes seem to move in three-dimensional patterns, trailing bright lines of digital 'energy' behind them as they mutate and merge, accompanied by electronic sounds of humming and beeping. At the start of *Sulphur 16* an oversized computer-generated projection of a ghostly body gradually emerges through a blue-purple wash then immediately fades out again. In both pieces, the live bodies of the dancers enter into this already-unreal technological environment; the uses of light, sound and projection and the ways in which the space is constructed all operate to remove us from human reality and transport us into a world in which time, space and the body seem to operate according to digital laws. Referring to these pieces, McGregor suggests that "the live body becomes the interface by which we experience and make sense of the technology" (McGregor 2000).

In *Nemesis* the process operates in reverse. It begins with a real-scale, two-dimensional, colour photograph projected onto the centre of the back wall of the stage space. This photograph, by Ravi Deepres, is of an empty room; a black and white tiled floor extends towards us from a large, heavy wooden door inset with greenish panels of frosted glass; to the left of the image, a white tiled dividing wall can be seen and the edge of a porcelain sink; the accompanying soundtrack incorporates suggestions of echoing and water dripping. What we are presented with is a mimetic photographic depiction of a public lavatory in some Victorian municipal building. Within the conventional theatrical discourse, this concrete image of a 'real' human environment positioned behind and above the stage floor suggests a setting to its audience. When two dancers walk onto the white-lit stage in front of it we relate them to it, which is to say we assume that their actions take place within it. *Nemesis* begins therefore by locating the performance within a place of familiar and measurable dimensions, a concrete space represented in human proportions. Throughout the first half of the piece it is the technology of this kind of space, the apparently real (or one might say the theatrical real), that McGregor explores and it is not until a detailed deconstruction of this scenic reality has taken place that the performance shifts into more overtly mediated 'virtual' spaces; even then, *Nemesis* sticks with relatively simple video imagery, eschewing the complex computer-generated projections of former pieces until the very end (when McGregor himself dances a final solo with a fluttering digital insect). In this return to more simple technologies of theatrical representation McGregor's intention is, I believe, to dismantle the dimensions of the real at their most fundamental spatial and temporal levels in order to discover and expose the virtual within them. By taking apart the 'hidden' technologies of the photograph and, thereby, deconstructing the notion of theatrical setting, he reveals the virtuality inherent in the performance experience and raises questions about the reality of the human experience to which it refers. In this case, to reverse McGregor's statement cited above, technology becomes the interface by which we experience and make sense of the live human body. It enables us to interrogate the seemingly natural relationship of that body to time and space. *Nemesis* is about the unreality of the 'human' real.

A major aspect of McGregor's purpose in the piece is to expose the conceptual dichotomies which already exist within the different environments of theatrical representation he employs. The discourse of the photograph for instance, conventionally constructs it as a factual documentation of reality but because it shows only a moment or fragment of that reality it has the effect of provoking an imaginary continuation: the viewer locates the meaning of the photograph in terms of a series of suppositions about what lies outside the frame. When it is used as a theatrical setting, these extra dimensions of the photograph are constructed by the audience in relation to the interaction between the photographic image and the other elements of the performance. In the perceptual hierarchy of dance, sound and light usually have a secondary importance. They are conventionally employed symbolically to create mood or atmosphere. In *Nemesis* however both sound and light have a powerful and intrusive presence. McGregor chose to work with sound-artist, Scanner, after hearing Scanner's 'sound violations', tracks derived from the covert sampling of the mobile-phone conversations of ordinary passers-by, distorted and played out live; McGregor was attracted by this invasive destruction of the boundaries between public and private space and the *Nemesis* soundtrack builds on this concept in a theatrical sense. Much of it is repetitive and discordant, creating a sense of interference rather than of harmony. Lucy Carter's lighting design centres around large-scale geometric projections which dominate the stage space and often seem either to trap or to overwhelm the dancers. Both sound and light swing between mimetic reflection of the reality proposed by the photographs and abstract counterpoint to it. In these ways, sound and light disrupt the coherence of the relationship between live movement and photographic setting, seeming to suggest alternative locations which the dancers simultaneously inhabit.

The importance given to visual and aural aspects of the production dislodges movement from its privileged position within the signification system of dance. Because of its abstract qualities dance is conventionally 'read' metaphorically or poetically rather than as a mimetic illustration of human activity but the presence of the living, breathing, sweating body on stage also accords it a capacity to achieve a direct sensual communication between performers and audience. Technology mediates the body, thus distancing the viewer from this sensual presence, but the capacity of the camera to zoom in close to bodies, to overcome the gap which exists between stage and audience in the theatrical environment, simultaneously allows it to create hyper-real representations. In the second half of the production, the live dancers are distanced from their own physical reality by the incorporation into their costumes of prosthetic pump-action limbs while behind them on the screen, large-scale video projections of their former, human, selves appear, including huge close-ups of their faces. Thus an uncertainty is created as to which is the more authentic of these two representations: the dancers are real because they are moving in the here and now, yet their futuristic costumes suggest a symbolic distance; the video projections are mediated, yet have more immediacy and presence because of their scale and 'humanity'.

McGregor has spoken of his desire to "work with a more digital aesthetic, to recreate an environment, an ecology of space which work[s] with the idea of a degenerating or electronic image, but [also] with a physical reality in a live space" (McGregor 1998). The importance of this statement lies in the fact that McGregor does not oppose the dancing body to the electronic image; he does not set the virtual up against the live, but rather speaks of 'an ecology', a word which conveys a sense of a relationship between the two where each gives something to and gains something from the other. The conventional discursive codes of each of the performance elements in *Nemesis* are destabilised both from within and in terms of their relationships with each other. Mimetic and symbolic structures of representation conflict and confuse our understanding of where presence and reality is to be located. In McGregor's words, "we provide a space between what the reality is and what the perceived reality is on stage" (McGregor 2003).

The idea of body

Speaking about his original conception of *Nemesis*, McGregor explains,

> My fundamental question was, 'what is body?' Could space have a body? Has time got a body? What is the feeling of body in time? How do you give light a sense of body? What is it that makes up an idea of body? That starting point, that interrogation of that starting point, has pulled out a whole new way of thinking. (McGregor 2003)

This statement offers us a clue as to the way in which McGregor's 'ecology of space' works. He is interested in the way our perceptual apparatuses operate between feeling and concept, between experience and idea. The environment or 'space' he wishes to create is one in which a dialogue can be enabled between the senses, the emotions and the intellect.

We can contextualise McGregor's notion of ecology in terms of recent theoretical models of performance and production. Ric Allsopp has proposed a notion of a 'performance ecology', where the use of the term ecology does not presuppose an essential and holistic connection between nature and art but instead suggests "a set of interdependencies between conceptual, social and environmental factors" (Allsopp 2000 4). In other words, nature and art are not uniquely and permanently related to each other by means of some single unifying discourse, one does not simply produce or express or reflect the other, but they do nevertheless interact. They do so at multiple points and in various ways. Within this ecology, Allsopp writes that performance operates as

> ...an unstable element or catalyst which enables people to rethink or revision or remediate their relationships to each other and to the interdependent worlds that are constituted by these relations. (Allsopp 2000 4)

It is clear that according to this conception reality is not simply a fact of nature but rather an aspect of experience which exists only in relation to other aspects. These relationships can be brought into question by performance.

Allsopp's model of performance ecology is comparable to Deleuze and Guattari's post-structural notion of the 'desiring-machine' (Deleuze and Guattari 1996). Desire, as Deleuze and Guattari express it, is a process: an ongoing, ever-changing flow. This flow is produced by the desiring-machine, a complex assemblage which operates in interaction with other complex assemblages in a continual process, each producing flows, which are then interrupted by other flows from other desiring-machines. In terms of performance, we can imagine the theatre as a kind of open-air marketplace into which a vast range of complex desirers bring their multiple desires to sell, buy and exchange. They intermingle and interact, deflecting, absorbing and transforming each other's flows of desire, raising and lowering the value of their product in relation to fluctuations in the movement of the whole. It is this ever-changing process of production with which Deleuze and Guattari are concerned, rather than with any singular product, which would constitute the end of, or reason for, the process.

This kind of ecology is neither stable nor systematic. There is no constant pattern of relationships which can be scientifically or objectively recovered in order to provide us with any one meaning or truth. The task of the analyst is to become aware of how different kinds of knowledge or experience flow from the various desiring machines which make up the performance. It is the interactions between desiring-machines which bring them to our attention. Allsopp's model shares with Deleuze and Guattari's a focus on the importance of breakdown. Deleuze and Guattari write,

In desiring-machines everything functions at the same time, but amid hiatuses and ruptures, breakdowns and failures, stalling and short-circuits, distances and fragmentations, within a sum that never succeeds in bringing its various parts together to form a whole. That is because the breaks in the process are productive and are reassemblies in and of themselves.

(Deleuze and Guattari 1996 42)

Deleuze and Guattari urge us to recognize 'lines of flight', flows of disruptive activity which operate between desiring-machines. Allsopp's model translates this idea into theatrical terms. He refers us to 'trauma' induced by some of the interactions which take place within performance. These are the moments of incomprehensible ambiguity where connections between elements seem arbitrary or contradictory. It is our resistance to incomprehensibility that forces us off the known path, onto other trajectories of comprehension.

Returning then to McGregor's 'idea of body', within the perceptual ecology he creates the body-concept must be multiple and continually in flux. Different meanings and values are formed, dismantled and re-formed in the interactive space between the various elements of the performance experience. In the case of *Nemesis*, the disruptive conflicts between various technologies produce a fragmented sense of time and space, within which the body loses its grounding. This does not mean however that the interactions between elements are random or unplanned. The choreographer describes his creative process as having three phases. The initial phase is conceptual and is undertaken with his collaborators[2]. McGregor comes with a set of ideas which he then expands in dialogue with them. The second phase is the generation of a "language, not just of the body but of all the other media" (McGregor 2003) and the final phase concerns composition. He explains what he means by this:

I think my responsibility as a choreographer is to work kinetically with every element, so I choreograph the film, I choreograph the lights. I work with choreographic form through all those processes. The collaborators all bring amazingly new things to the dialogue, but I do want them to look at their process from a choreographic point of view. So they're not coming in just bringing their own language. They're actually approaching it from what I call a choreographic consensus.

(McGregor 2003)

McGregor's desire to achieve consensus within his collaborative creative process does not indicate that he is aiming to produce a unified experience of performance for the audience. He is conscious of the various 'horizons of expectation' (Pavis 1982 74)[3] which are constructed by different media and that the interaction of these will create a disruption in the spectator's activity of meaning-making, making them "work harder at watching" (McGregor 2003). In their attempts to make sense of the performance the audience constantly experience deflections of their interpretative flow which make them conscious of the different languages in play. Their idea of the meaning of the performance therefore emerges from a process of continuous active negotiation. They must, to paraphrase Allsopp, rethink, revision or remediate their 'natural' perceptual pathways.

Refreshing dimensions

The thing about reality is that it's synchronous and multiple. My work is like that and people think that's really strange, but actually that's how we live. On the internet, you lose yourself in time and you follow through a kind of mind-map as you go from one site to another. It can take you in very bizarre directions. You arrive at a place and you think, 'how did I get here?' I think about time like that. What I want to do is capture an element of that. The intention is not to be confusing or incoherent, I'm dealing with reality.
(McGregor 2003)

The time-space of *Nemesis* is analogous to the digital time-space described above. It is constructed like a series of screens, one laid over the top of another yet all remaining present, so that the spectator sees the trace of one through the other. This occurs in terms of the structure of the whole piece, in that the more overtly 'technological' second half operates on some levels as a 'revisioning' of the first half (a point to which I will return later), but it also occurs within each separate sequence of the first half. The photographs projected in the first half of the piece initially suggest a conventional mimetic setting for the dance, a location within which the stage action can be supposed to be taking place in 'real' time. In one early sequence an image of an empty, dilapidated room with a fireplace at the rear is projected. In this photograph a pattern of light falls on to the floor as if through a window just outside the frame on the right. This same pattern of light, slightly larger in scale, is seen on the stage floor, also on the right, so that the dancers can be supposed to be moving within the room we see in the photograph. While the photograph remains, however, the lighting changes. The 'window' light disappears and is replaced by three rectangles of light on the left of the stage floor with a veined pattern running through them. Each of the three dancers now inhabits their own rectangle as if caught within it. This veined pattern of light then spreads across the whole stage floor and eventually across the back wall too, swallowing up the photograph and dwarfing the dancers. The 'idea of body' is thus problematised in terms of its relation to spatial dimensions.

Because the initial relationship between the photograph and the light is mimetic the spectator at first comprehends the dancers' location in terms of the concrete reality indicated by their interaction. Even as the light changes we cling to the idea that the dancers remain somehow 'within' the room. I interpreted the three veined rectangles as magnifications, as if a camera had zoomed in on the sunlight cast through the window onto the floor to reveal trails of dirt in cracks in the linoleum. The spread of the pattern then seemed to me to suggest a further magnification. In this context as the space expanded, by logical extension the dancers appeared to have shrunk, to have become, in effect, some other microscopic form of life normally invisible to the naked eye.

A later sequence operates in reverse perspective. A photograph which seems to be only an abstract pattern of squares is projected onto the back wall and is mimicked in light by a grid shape on the stage floor. In this instance since we cannot 'read' the photographic image mimetically the initial interpretation of the dance's location is dictated by the pattern of light and is more symbolic than literal. The dancers seem to be trapped within this rigid, cage-like grid. As the sequence develops however a series of further shots appear, as if the camera is gradually pulling back. Now it can be seen that the original photograph was actually an extreme close-up of a cushioned seat within yet another empty room. The lighting reflects this transformation so that we realise that it was, all along, a deictic but magnified representation of the room. In reflecting on the whole sequence we realise that the dancers have 'grown' from microscopic inhabitants of the seat to human-size inhabitants of the room.

In both examples, McGregor carries his audience back and forth through the spatial dimensions revealing, as he does so, the flexibility and contingency of spatial perception. In Deleuze and Guattari's terms, the 'line of flight' here is produced by the breakdown in the 'natural' dimensions: because the spatial and temporal dimensions of the theatrical reality are unstable, the body also loses its location. Unsure then of 'where', 'when' or 'what' we are watching, we hold in mind a series of different pictures, each representative of a different, but simultaneously existing reality. McGregor has spoken of his fascination with the unseen in any space, the fact that what you see is not the whole picture (Random 2002 video). What the examples above illustrate is the way in which, in *Nemesis*, he excavates moments in space and time, peeling back layers of perception in a search for further dimensions of the present event: "You put light and body and graphics together to expose a moment, to elongate a moment, to backtrack in a moment. You enable the audience to experience time as fractured" (McGregor 2003).

Through his exploration of the technology of the photograph, its capacity to magnify and distort, its ability to transform and multiply our perception of a space by altering the perspective from which we experience it, McGregor succeeds in revealing the activity of imaginary levels of perception within even the simplest dimensions of the real. He demonstrates how our basic perceptions of breadth, height and depth operate in continual interaction with less absolute notions of space and time to create complex imaginary locations. Trusting initially in the actuality of the photographic rooms, we gradually gain a sense that they are not 'all there is' and an awareness of our own activity of filling in the gaps through imaginative projection. It becomes clear that our vision is working in constant interaction with a range of other perceptual dimensions to continually construct, dismantle and reconstruct what we imagine we see.

The fiction of the virtual

The first half of *Nemesis* is constructed out of a series of disconnected 'excavated moments' such as those described above. Each scenic 'room' is separate from the last, containing its own distinct (and disjunctive) patterns of light, sound and movement. There is nothing to suggest that these are all rooms in any one building and there is no sense of an overall narrative progression, rather the piece presents itself as a collection of unstable frames, randomly structured. The only continuity in fact, is a thematic one of breakdown in the relationship between body, space and time. The theatrical 'reality' presented, then, is a fragmented and uncertain one. It is only once our sense of the stability of the 'real' dimensions of space-time has been thoroughly dismantled in this way that McGregor takes his audience into a properly 'virtual' world.

Nemesis is composed of two apparently quite different sections, linked by a transition which transforms the concrete reality of the photograph into the virtual reality of video. Midway through the piece, the dancers are moving in front of a projected photograph of a fireplace. A flickering, fire-like lighting effect consumes first this image and then the whole stage space, as if to suggest the destruction of the concrete world to which the performance has so far referred us. Supporting this interpretation of destruction, both music and movement build to a kind of chaotic climax. The lighting effect then turns from red-yellow to blue-white, like interference on a television screen (an impression encouraged by a kind of electronic blurring over the music in the soundtrack) and finally resolves itself into the original projected photographic image of the tiled cloakroom we saw at the beginning of the production. A single dancer is left on stage. She watches as a 'virtual' dancer appears simultaneously within the photographic space. As the real dancer exits the stage space, her virtual image crosses the projected space, which dissolves as she moves and is reconstituted into a new room, a video image this time, projected onto the screen. Static, semi-transparent video images of other dancers repeatedly fade up and fade out 'within' this screen room, sometimes layered one over another. The effect is like that of a series of shots taken by a time-lapse camera.

What is interesting about this space is that it is, on the one hand, akin to the photographic settings we have seen throughout the first half (in that it is another real-scale representation of an empty room), but on the other hand, less real because it is a video projection and we conventionally think of the time-space 'rules' of video as being more flexible than those of the photograph, at least when it is used as theatrical setting as it has been here. To clarify, at no point in the first half did the dancers actually appear within the photograph as they appear now within the video, yet their transportation into this 'virtual' room reminds us that we have been imagining them as being 'inside' the rooms all along. It is as if our imaginary perception that the dancers inhabited those photographic spaces has been actualised, as if we ourselves, through our activity of interpretation, have transported those 'live' bodies into this 'virtual' space. Here is an example of the way in which Deleuze and Guattari's deflections of flow function to reveal the operation of the desiring machines which produce flows. Our interpretative habits become present to us in this moment. The dimensions of space and time which constitute our conventional understanding of reality (represented by the photographs of 'real' rooms) have already been destabilised, expanded and complicated so that this virtual world seems less wholly 'other' and more like just one more manifestation of the real-imaginary, equally flexible and contingent, no more or less constructed than our interpretations of the previous 'live' action.

After a few moments, the video projection zooms in on the blank and humming screen of a television which then grows to encompass the whole stage and screen space in a flickering light. This virtual action is accompanied by a soundtrack which suggests interference, as if someone is searching for the correct frequency on a radio. Jumping, blue images of rooms we have seen before in the photographs appear transiently before us, now resembling television images. They bleed off the screen onto the stage floor, as if cannibalising the live presence of this stage space, subsuming it into a virtual, mediated universe. The theatrical space is hence redesignated as virtual, or rather, the two environments of stage and screen become interchangeable, neither being in possession of a more concrete reality than the other. Right up at the back of the stage, where it casts a large shadow onto the screen, a strange moving form appears, human in size and shape but with one elongated limb and dressed in a dark shiny costume reminiscent both of a beetle and of a futuristic space-creature. The figure goes through a kind of insect-like process of metamorphosis, a sequence of jerky movements during which it seems to be gradually acquiring control over its own body. The implication is that this new technological world has given birth to a new sort of technological creature, perhaps human, perhaps not.

In terms of performance discourse, the two halves of *Nemesis* are simultaneously different and alike. Through the transition described above, we understand that we have moved from a theatrical environment to a televisual one: the accompanying metamorphosis of the 'human' dancers into futuristic, technological creatures has the effect of interrupting our supposed sensual connection with their bodies, diluting their 'liveness' and making them seem more 'fictional', rather like cartoon animations. At points however, large-scale projected images of the dancers' faces and bodies appear in hyper-real close-up behind themselves on the screen. The human body is here 'doubled' in time and space yet the screen images look more 'human' than the moving bodies present on the stage. This aspect of the production, together with spatial and kinetic familiarities in the choreography of movement, image, light and sound suggest that the second half might be, to paraphrase Allsopp, some kind of distorted 'rethinking or revisioning or remediation' of the first. A dialogue is set up between the systems of the live and the virtual where the codifiers which make up our notions of time, space and body within those systems are displaced. If we return to Deleuze and Guattari's notion of the desiring machine we can see that this dialogue reconstructs both the virtual and the real as unstable and interactive complex assemblages, each system dependent on the other for its identity.

Figure 1. *Nemesis*, Sadler's Wells Theatre, 2002. (© Ravi Deepres).

Body-technology

Deleuze and Guattari urge us to seek, in analysis, the 'non-human' (Deleuze and Guattari 1996). In his introduction to *Anti-Oedipus*, Mark Seem explains this in the following terms:

> *They urge mankind to strip itself of all anthropomorphic and anthropological armouring, all myth and tragedy and all existentialism, in order to perceive what is non-human in man, his will and his forces, his transformations and mutations. (Deleuze and Guattari 1996 xx)*

Basically, what Deleuze and Guattari are suggesting is that we attempt to displace the concept 'human' from the centre of our vision, that we cease to comprehend everything in terms of the human discourse, in order to attain a more 'ecological' perspective which enables us to see 'man' not as a stable and defined force but as mutable and interactive in relation to those other forces which make up the world.

McGregor has described the body as being 'inherently technological' and in his earlier work he has sought to "extend its capability with the intervention of media devices" (McGregor 2000). Previously though his strategy has been to project computerised images of the extended or re-visioned body with which real bodies then interact. In *Nemesis*, the dancers' second half costumes include pump action prosthetic arms, so that they themselves begin to function technologically. It is only in this second section of the production, when we see the dancers moving with the prosthetic limbs, that we gain a kinetic understanding of the movement vocabulary of the whole piece. It becomes apparent that much of the choreography of the first section has in part developed from the company's work with these limbs. McGregor describes the process:

I was very interested in the whole idea of artificiality, and how artificial extensions change the way you look at bodies. It is incredible how just that extension changes everything, from the centre of gravity to where your rotation is to having to work different muscles in your arms, how it affects the back...It had such an impact on the way in which we started to think about movement ...You unlearn what the technology of the body does. The dancers had to unlearn how to turn, how to hold someone else, how to use the full circular extension. They now have that physical information in their bodies but the prostheses aren't there any more. That experience of a different physicality is very interesting (McGregor 2003)

McGregor's movement language in *Nemesis* plays a kind of double game around notions of the organic and the technological, travelling between extremes of incapacity and superhuman flexibility and strength. In the first half, the dancers' bodies seem to stretch and twist themselves almost to a punishing degree but their exertions are interrupted by momentary breakdowns, glitches in the muscular technology. On one level they seem extra-human, practising levels of control and discipline which seem excessive; but in relation to the sound and lighting designs and the photographic tricks of perspective, that excess can sometimes take on a different feel. When the dancers are dwarfed by the design, as in the two sequences from the first half that I described earlier, their movement is read in relation to the location the setting accords them. When the window-light mutates into the veined pattern and fills the stage and screen the dancers are transformed into microcosmic creatures and the jumpiness and lack of flow in their movement evokes associations with insects. When they are dancing within the grid lighting pattern, this image of insects takes on a further level of poetic meaning; the geometric shape and mechanical, repetitive sound make me think of technological bugs, computer viruses. In the realistic setting and intimate duets and trios of the first half the jerky halts and strange twists of the body imply personal breakdown or collapse. In the second half those movements are transformed by the change in costume. In their new *Star Wars* aspect, flourishing their prosthetic limbs like weapons, the dancers resemble futuristic superheroes.

Yet the metamorphosis remains incomplete in several important ways: in the live production I saw, there were moments, small incidences of imbalance or missed rhythm, where it was clear that the dancers were not yet fully attuned to their extended limbs. These momentary breakdowns interrupted the fictional world created to remind me that there were human bodies inside the technological costumes, which were in some way disabled by their technological additions. They also revealed to me the choreographic connections between the first and second halves: *Nemesis* dislodges the body from its position within dance discourse as central coordinate of the human time-space map; the small struggles the dancers undergo in the second half to deal with their newly extended body-space and body-time are the conceptual point from which those glitches and hesitations in the movement language of the first half have emerged. The effect of the prosthetic limbs then, works in conjunction with the video projections of the 'real' dancers I mentioned above to create a series of perceptual dichotomies between ability and disability, the human and the non-human, live presence and technological mediation, reality and fiction, which challenge the conventional perceptual economy of dance. The dancers in *Nemesis* are not one thing, but many; the production obscures the clarity of the body which lies at the heart of contemporary dance discourse, representing it as a fragmented and multiple entity in constant tension with itself. The piece also dislocates the spectator's sensual and emotional identification with the live. As the dancers 'unlearn' how to move, the audience 'unlearns' how to understand dance.

The ecology of perception

One of the simplest yet most radical compositional strategies McGregor employs is to disrupt the conventional hierarchy of theatrical signifying systems. In Random's work, the body is not the privileged term as is normally the case in dance. Instead it fights for space with all the other elements. It is within this complex, disjunctive interaction between facets of the theatrical experience that the operation of McGregor's digital ecology can best be realised. In describing his interactions with his collaborators, McGregor repeatedly refers to a notion of creative tension. The inspiration for Scanner's original score lies in sampled mobile phone conversations. This is an approach which fascinates McGregor:

> It's about public and private space, a spatial concept in sound. These are private conversations, some of them very intimate, someone's having a really awful argument that's sampled and extended into music. Something in the sensibility of that, the difficulty of it or the tension in it gives you a really fantastic sound world in which to trace choreography. Physically we worked a lot with the tone and the sensibility of that music, the physical translation of it in the body. (McGregor 2003)

He offers a similar account of his interaction with photographer and digital video designer, Ravi Deepres:

> Ravi would decide how we might work with the idea of the photograph, then I'd work with him in terms of what that means about plane, structure, order, all those fundamental things. …It's fantastically challenging because of course his visual sensibility, his visual intelligence or literacy is very different from mine, and that tension provides a very interesting conversation on stage. (McGregor 2003)

The creative spur McGregor discovers within the tension between various economies of representation refers us back to Deleuze and Guattari's 'lines of flight' or Ric Allsopp's notion of interpretative 'trauma'. The audience experience *Nemesis* as a series of ambiguities and contradictions which shift them back and forth between different perceptual pathways. Nothing is more real to us than physical presence but McGregor reveals to his audience that presence is itself a constructed concept. His audience must constantly reconfigure their idea of body within spatial and temporal dimensions which are perpetually in flux.

In McGregor's ecology of space, the real is an interactive process without an end product. Ecology does not suggest progress in the sense of a linear journey towards a distantly realised ideal, but is haphazard, nomadic and multiplex. The production achieves a sense of the complexity of experience: the simultaneous existence of past memory and future projection within the temporal present; the microcosmic and macrocosmic potential of the spatial present; the interaction of the physical, emotional, sensual and imaginative which make up our perception of corporeal presence. Technology, for McGregor, is a means of showing us reality:

> I think that my work is more real than a lot of work that is supposedly reality-based. What's really interesting to me is that the ambition in so much virtuality is to create as 'lifelike' an experience as possible, yet the experience of life when you're living it is very different: it's synchronous, it's multipilicit. That's all I show in the work and people often think that's strange, but actually that's how we live.
> (McGregor 2003)

This final statement points us towards the wider relevance of McGregor's work. There is a deep anxiety about the use of technology which is perhaps most apparent in the dance world but is also present in theatre generally. The fear is that, if we 'lose' the body, we lose some essence of human experience, and thereby divest physical performance of its essential relevance and power. McGregor takes the opposite view, believing that a more complex exploration of technology can give us access to a broader and fuller understanding of human experience. He suggests that the digital universe is no more technological, no more 'virtual' than the theatrical universe, or indeed, than the human 'real'. In giving equal weight to all aspects of the production within his choreographic process, he demonstrates that the 'idea of body', is just that, an idea, and one which the discourses of theatrical and artistic representation have always felt free to play with and manipulate. His 'digital aesthetic' merely makes that manipulation overt. Other dance and physical theatre companies, such as DV8 and Theatre de Complicite, have also moved towards highly technological, multi-media productions, not so much leaving behind them the sensual reality of corporeal experience as accepting that it is only one reality among others. Increasingly, theatre practitioners see technology as a necessary tool for 'realistic' representation of contemporary experience. To me, this approach implies a liberation of perspective, a welcome call away from absolute notions of reality towards a more multi-faceted and interactive experience of the many simultaneous differences of our world.

Notes

[1] See Schechner, R. (1988) *Performance theory*. Routledge, London. Chapter 1, pp. 1–34.

[2] For *Nemesis*: original score by Scanner; animatronics by Jim Henson's Creature Workshop; photography and digital video design by Ravi Deepres; lighting design by Lucy Carter.

[3] Pavis borrows this term from H. R. Jauss who explains that the public for a work is "predisposed to a certain mode of reception by an interplay of messages, signals—manifest or latent—of implicit references and of characteristics which are already familiar". Each new text "evokes for the reader…the expectation horizon and the rules of the game with which preceding texts have made him familiar".

References

Allsopp, R. (2000) The location and dislocation of theatre. *Performance Research* 5(1) 1-8.

Deleuze, G. and Guattari, F. (1996) *Anti-Oedipus: capitalism and schizophrenia*. Athlone Press, London.

McGregor, W. (1998) On the making of Millenarium: Wayne McGregor in conversation with Jo Butterworth. In Butterworth, J. and Clark, G. (eds.) *Dancemakers in conversation*. Bretton Hall College, University of Leeds. http://www.randomdance.org/random/index.html. Accessed March 2003.

McGregor, W. (2000) Dialoguing ecology. *Animated Magazine* Autumn 2000. http://www.randomdance.org/random/index.html. Accessed March 2003.

McGregor, W. (2003) Unpublished interview with the author, March 17th, 2003, London.

Pavis, P. (1982) *Languages of the stage: essays in the semiology of theatre*. PAJ Publications, Baltimore.

Random Dance (2002) *Nemesis*. Sadler's Wells, London.

Random Dance (2002) *Nemesis* unbroadcast video. In association with BBC4.

6 halving angels: technology's poem

Jools Gilson-Ellis and Richard Povall

Technology has made different kinds of poets out of us. Together we sing ghost songs. We have haunted mouths, and speaking flesh. Together we imagine impossible things that I can write, but not make. Together we make things that I can't imagine. We barter noisily like grandmothers. Because I am a writer, and trade in poetry, so I tempt technology to do the same. (Jools Gilson-Ellis)

This article uses *halfangel*'s art and performance practice to analyse the ways in which digital technologies have altered processes of making work. It suggests that technology's impact might be in the poetic transformation of imagination. The article focuses in particular on *halfangel*'s 2002 work, *Spinstren*[1].

halfangel is a performance production company based in Ireland and England. The company was formed in 1995 by Jools Gilson-Ellis and Richard Povall. *halfangel* has developed a distinctive body of work through the long term interdisciplinary partnership of its co-directors. This is characterised by a poetic use of new and emerging technologies, with platforms including CD-ROM, installation and performance. Povall's work as a sound and video artist includes the development of intelligent performance environments, closely tuned to the poetic gesture of Gilson-Ellis's writing. Gilson-Ellis is a choreographer and writer, with a strong sense of the poetic possibilities of emerging technologies. Their work is notable for its unsettling use of voice, movement and sound.

halfangel's CD-ROM *mouthplace* (1997) was exhibited at international festivals in five countries. The work developed the troubling connections between femininity and orality.

The dance theatre production *The Secret Project* (1999) was co-produced by The Banff Centre for the Arts (Canada) and the Institute for Choreography & Dance (Ireland), and was performed in four countries. *The Secret Project* implemented several years of performance research at The Banff Centre, and focused on the idea of the 'secret'. Our recent work *Spinstren* (2002) is an exploration of the metaphor of spinning. It combines new and old technologies— the spinning of thread as well as voices. *Spinstren* was premiered in Ireland and England in Spring 2002, and will tour during 2003.

*

JGE: Why would you want to halve an angel? The name of our company comes from a trapeze move I once learnt from the trapeze artist and teacher Deborah Pope. A half-angel involves sitting on the trapeze and holding on to the bar with one hand, flexing the foot of the opposite leg, and falling backwards. If it works, you end up hanging beneath the trapeze from a grasped hand and a flexed foot, and you can extend the free arm and leg in an elegant diagonal. If it works, you are half an angel. I loved learning trapeze, and this move in particular, because it involved the breathless moment between falling and catching. One moment you are sitting prettily on a trapeze like a girl on a swing, the next moment you fall backwards, with at least one hand not holding on. It is a magic trick. Something of it caught me, just as I am caught by a sturdy flexed foot, and a grasped palm. In our work, I have been repeatedly drawn to the idea of falling as a moment when conscious control is lost, and other imaginative possibilities might approach. It makes sense, then, that our angel isn't whole. She's only half an angel, and she shifts oddly between flying and falling. For a long time we thought there would be angels in *The Secret Project*; glimpses of creatures with wings seen in the half-light, until one day of revelation in Banff, when all of the angels got pulled off the wall, and all that remained was the single spoken word that opened the piece; an understated breath saying 'angel'.

RP: Actually, this isn't true. Most of the angels got ripped off the wall, and there is the understated breath of 'angel' that begins the piece. But one of the angel cards[2] remained; a duet between Cindy and Mary[3] in which they trade off word and gesture. We made a list of words that might precede 'angel' (broken angel, weeping angel, giggling angel, sorry angel…) and a list of words that might come after 'angel' (angel cake, angel wings, angel fish, angel eyes…). In performance, Cindy and Mary stand beside each other, down and centre stage. Mary speaks the word 'angel', which causes our system to choose randomly a recorded word to accompany it (Mary: angel…cake); Cindy moves—a gesture in sympathy with angel cakes. Mary responds with 'angel' again, which in turn triggers the next word. These are half angels, and a beginning of our tentative moves towards using generated text in our work (see below). How interesting that this should have been forgotten…

JGE: In the trapeze move of the half-angel, falling becomes flight. Or to put it more precisely, one can only fly if one is prepared to fall. In falling, there is always the possibility of flight. For me, femininity is entwined with these possibilities of flight and falling. Angels, of course, have the same problem. Perhaps the only way to broach what might appear as a contradiction between flying and falling, is to steal its gesture, somewhere in the weary oscillation between one thing and the other. After all, in French, flying and stealing are the same thing[4].

Enchantment. Here is a story about a girl who steals a top. A wooden top, precious to its owner, who had carved it by hand. This is a story about a girl who fell in love with spinning, not because of movement, but because of

stillness. This is a story about a girl who stole spinning, because she thought she could possess it. This is a story about a girl who learnt a difficult difference between stealing stillness and stealing movement. This is a story about a girl.
('Enchantment' from Spinstren, *Gilson-Ellis 2002*)
Later, she sits in the valley café drinking tea, and Joe comes in & moves into the booth across from her. 'I'm not falling' she says. 'I know' he says. 'It's snow ghosts, tripping me'. 'I know' he says & turns to order coffee.
('Snow Ghosts' from The Secret Project, *Gilson-Ellis, 1999*)

Half-angel to *halfʃangel*. If we were an art angel, or the third angel, or a company of angels[5], I suspect we would fly differently. But we are *halfʃangel*. We hang a little precariously from a flexed foot and a sweaty hand. If angels function as a signatory to hope, then I hope for a stolen falling flight. Such a flight is the poetically-conceived space of our work. Such a flight is composed of a wide tangle of poetic text, sound, performance and technology. Such a flight is a poesis, whether its constituents are actual poetry or not. And they are not, as often as they are.

RP: We build poetic systems. They are systems that, crudely speaking, capture the essence of the performing body, turning that information into sonic or visual gesture. In these systems, the centre of attention is not the objective body-state. The focus instead is the subjective body—the phenomenal dancer. I use off-the-shelf video-based motion sensing systems that use a video camera to watch the dancing body(ies) and a computer to analyse that movement and turn it into a sonic or visual gesture. For many years, we used Steim's *BigEye*, but more recently have switched to using David Rokeby's *SoftVNS*. Both of these systems are highly extensible, and allow me great latitude in re-making their purpose. These kinds of systems are designed to provide objective (empiric) information about the movement of a body (or any object), but we don't use them that way. Instead, we build phenomenally-aware systems that look at the phenomenal dancer rather than the empirical dancer: an observation of how she is moving, an attempt to extrapolate the reasons why and how she is moving—an attempt to create a phenomenal understanding of the movement. Why do we do this? What is it that we found lacking in the objective data about the moving body? In the early research, something just didn't feel right:

…Science succeeds in constructing only a semblance of subjectivity: it introduces sensations which are things, just where experience shows that there are meaningful patterns; it forces the phenomenal universe into categories which make sense only in the universe of science. [It cannot realise] that the perceived, by its nature, admits of the ambiguous, the shifting, and is shaped by its content.
(Merleau-Ponty 1964 11)

We found, almost by accident (when Jools was dancing holding a 3m. length of bright blue electrical conduit purloined from the streets of Amsterdam in each hand) that, rather than tracking literal body movement, we could begin to look at the information phenomenologically, looking instead at essence. In this case, the camera was told to see only the tips of the prosthetic conduit arms—and we saw something magical. When we began to trigger textual gestures with physical gestures, something alchemical had happened. In creating a system that can extend physical gesture into textual gesture, the aesthetics of movement become the aesthetics of an emergent poetic narrative. Media artist Bill Seaman talks about emergent meaning and recombinant poetics, where "meaning emerges through combinatorial relations of differing media elements"[6]. He describes this as "a new language, a new approach to linguistics". These systems of ours were not arrived at overnight, but after years of experimentation; years of finding a way in to the technology that will allow it to find a poetic voice of its own.

JGE: In Jeanette Winterson's novel *Sexing the Cherry* (1987), the main character Jordan goes to a town where all the phrases, curses, and sweet nothings uttered by the inhabitants float up and clutter the sky with their cyclical chatter. Jordan goes up in a balloon with a woman whose job it is to mop the sky of such verbal ephemera. Although she isn't supposed to, this woman gives Jordan a small box, in which is caught a sonnet. If he opens the box carefully, he can hear it repeating endlessly (Winterson 1987 18). I've loved Jeanette Winterson's work for a decade and a half. What *halfangel* does is make the box into a space that you can dance inside, so that Winterson's disembodied vocal traces become oddly connected to moving bodies. Such bodies are ghostly in their relation to this utterance. Sometimes it seems clear that they speak their poem before us, but at other times, we are troubled by the nature of the connection between flesh and sound. As she moves, so phrases tumble. If she is still, there is often silence. Sometimes when she moves and speaks, there are too many voices for one body. It might be that she shifts in memory, repeating something she hears tumbling. Or perhaps she sets the poem-spell running with a different kind of agency. She navigates time oddly. Something of the grammar of her moving body is in the present tense (she is here before us), but because the voices are sometimes visibly uttered by her, and at other times 'moved' by her (the same voice, the same poem, but connected to / triggered by movement), utterance troubles time[7]. Jordan's sonnet repeats endlessly, curling back on itself. With our sonnet, the navigation between the present tense of voicing and moving, and the past tense of writing and recorded voice, perform another kind of resistance. This is a resistance to the difference between past and present; a resistance to the difference between internal and external worlds. To confuse the two of these in performance is an unsettling enchantment. When it works well, I believe it approaches the operations of the unconscious, and it is this that is so troubling, both to perform and to watch.

My writing always had a bent for the magical, but the experience of working with technologies in this way makes me a different kind of writer. If I was irreverent about technology's limitations when I had no knowledge of them, now I imagine vast impossible alchemies that make Richard giggle. What I want to suggest is that this is important. Such an alchemical gesture is rooted in a physical understanding of the poetics of the work. But it also always asks, 'What could we make if we were better magicians?' This writing is part of a process in which Richard and I demand a poetic alchemy from the digital.

Figure 1.
Jools Gilson-Ellis in half/angel's *Spinstren*. © half/angel 2001

This gesture of challenging technology's digital duplicity into the trouble of poetry's metonym (in whatever genre) is a double trick[8]. We do not want to see technology, we want always to see its spell.

RP: Of course, I'd worked with others who wanted to work with technology but not within it (no alchemical processes allowed, sorry). Their demands made me inwardly sigh, but not giggle. If Jools has impossible dreams about how we can use our technologies, they are dreams that still retain a place in the final work: a shared dreaming space, perhaps. The world behind the computer screen is a pragmatic one; my job is to let poetic spells weave their magic to overcome the limitations and realities of dumb hardware and pre-cogniscent software. I build systems that have an intrinsic poetic space of their own—sonic spaces that yearn; spaces that are obsessively languorous, unsettling an audience and disconnecting them from reality. In *Spinstren*, the three-dimensional soundworld hovers between the real, the remembered, and the imagined. Visual imagery weaves its own spell, in its own time, as well as improvising with text grammars to create new narratives, barely glimpsed.

*

The three-dimensional soundworld we use in *Spinstren* is a special haunting within the physical performance space. I am using specific techniques and technologies (none of which I can claim to have invented) to expand the perception of physical space. I can send sound spinning around within this three-dimensional space, sometimes gently and slowly, at other times impossibly fast. The soundworld underpins the entire piece—there are very few moments when we are not within a sound environment. The demands of poetic alchemy are sometimes those which simply call for a light touch, for a technoetic (Ascott 1999) space which is tender and peripheral.

JGE: After the show, I introduce Richard to Maud. 'This is the composer' I say. 'Ah!' she says, 'you're the one who makes it float.'

*

Underworld

> Carla is caught staring at a thousand webs in the field beside the well house. Carla is caught. Where the field is, there is an underworld of tiny spiders, who rise to the surface, and weave another fabric to lay across the grass. She knows that if the light shifts she'll not see it again. So she stands stock still, with her back against the well house door. In wonder. Carla is caught.
> ('Underworld' from Spinstren, Gilson-Ellis 2002)

JGE: This is something I saw. An impossible thing. I run breathless to the cottage to get the camera. Back at the well house, I crouch on my haunches with the video poked through the fence. It's so magical, it doesn't look real. A field of iridescent webs, moving like an ocean in sunlight. I am enchanted it is captured on video at all. Richard edits it simply, overlaying the text. After the show, people ask us what effect we put on the grass to make it shimmer so. Technology can make wonder, but it can also steal it.

RP: Video, after all, is all-too-often an all-too-literal space, except when it descends into a postmodern, distressed fog or an undergraduate abstraction. Working with videospace in realbodyspace is a challenge that few have met (we feel) with success. Videospace is a dominant space, that grabs and unseats the magical and which competes for our attention, demanding it without compromise or recompense. The lonely body(-ies) on the stage look forlornly on as the audience's eye is stolen by gleaming photons. We have experimented with combining the realbodyspace with the videospace, but have not yet had the courage to take our experiments to performance. Instead, I'm always insistent on using real images (never computer generated, rarely digitally-invented or composited) to create the unreal. Sometimes words are projected (in the case of *The Secret Project*, onto a curtain of falling rice). Moments of beauty prick the continuance of body-narrative, tell their story, and fade away.

The Lost Song

JGE: 'The Lost Song' was one of many texts I wrote for *Spinstren* which were not used in the final performance. But this text is different. It's different because, in another sense, it was used. I liked it as a text, because it tried to wrestle with what it meant to sing, but also what it meant to be haunted by song. I had this idea that Carla was incompletely taken by these groups of women called the *Spinstren*. The *Spinstren*, who use spinning tops to weave all kinds of healing and troubling spells, aren't sure what happened with Carla. The top Carla stole was one of their hunting tops. These are knowing and gentle hunters, and the compulsion to steal them happens only when a choice is made about becoming one of the *Spinstren*. Such a choosing is woven thing, not something made clearly by either Carla or the *Spinstren*, but a moment when she and they and the top span perfectly, and the unutterable balance between falling and spinning and stillness compelled Carla to take the top. Other (rare) times when the *Spinstren* hunting tops were stolen, there followed a slow process of transferral; something I can't tell you about because it is a deep secret. The hunting tops didn't hunt for new recruits, they hunted for a moment when time and spinning and falling came together powerfully. In this way, those who stole the tops, were already moving towards the *Spinstren*. This was how it was with all of them. But not Carla. Carla was caught between worlds; an unheard of trouble. And the *Spinstren* were troubled. They could not bring her to them, nor could they release her. But they could sing to her:

> *Carla woke up singing. It had happened twice before, but only since she stole the damn top. When they move towards her, one beside the other, tall and stout and slight, their woven voices in a four-thread, their singing moved inside her, lifting the lost song. This was the song, the lost song. And as they moved past her, she wept. And then there was nothing.*
> *Here in the roar of it came a moving steady knowledge. She let herself loose from conscious shores, and belly sound wound from her. It moved around and beside and inside the song between them. Unknowingly, she knew where her*

voice could find its way between the other ribbons of song. Sometimes their voices rose to a tumult, all of them keeping voices rising and swelling in strange harmonies. Other times she would turn and find herself in a whispering conspirational tangle—intimate and urgent. Carla woke out of breath.

They were calling her. She could hear it in her deep sleep. Gentle almost— sounds, in harmony.

('The Lost Song' from Spinstren, *Jools Gilson-Ellis, 2002*)

For some reason, this text faltered in rehearsal. Eventually I realised that this was because what I wanted was singing, not a description of singing. Cindy, who has a knack for these things, had been working out how to use the peg tops. These are old-fashioned wooden tops wound around with string, which you throw in order to spin. It's extraordinary when it works, because it flies from your hand through the air, and lands spinning. We went back to 'The Lost Song' and began to work with its sensibility. What would it be like to have lost a song? How might you sing yourself towards it? We sang with the peg top clasped in our hands. We played with traces of lost songs, and then we would throw the top, and as it hit the stage and (hopefully) span, so we would move into a full-bodied song, and as the top slowly ceased to spin and toppled, so our song would stop. Richard watched all this, and suggested we try to design an intelligent space in which we could literally find traces of a lost song in the air. So we did. And so Richard made a space in which I am lost in a forest of half-heard voices. And just as I described before with poetry, this is a space in which times and corporeal borders become muddled.

RP: How do I make space intelligent?! (laugh) Artificial Intelligence mumbo-jumbo aside, I want to make the case that these systems are emotionally intelligent because they sense phenomenally. Here we wanted to evoke the loss of the song—a troubled kind of finding. I asked Jools to sing, to give me the material for a library of sung phrases; some were extremely short, barely vocalised, others were longer. All were tentative, often placed on the edge of voicing. These I split into two classes—the tiny barely vocal phrases, and the rest, more clearly sung. The system is designed to pick up tiny movement, almost as invisible to the audience as to the computer's eye. The anchor for this loss is melancholy; it isn't a maddened chase. Jools is rooted in the floor, so almost all of the movement comes from the upper body. The lighting is very dim, so the system is pushed to its limit as it tries to see through the gloom[9]. The system is relatively simple. I use a counter to expand the possible library of sounds that could be triggered. The first set of sounds are only from the first class, as Jools attempts to find her voice; later the more clearly vocalised phrases begin to appear, as Jools begins to inhabit her voice and the song. I also add a long sustained open fifth vocal chord—a sonic chair...

If Merleau-Ponty suggests that consciousness is a fundamental property of a phenomenal system (1964 58–9), then in creating a phenomenal motion sensing system, are we implying that the machine itself has consciousness? In one sense, yes (phenomenally), but at the objective level, it's patently rubbish. So what is happening is that we are introducing phenomenological understanding by the way in which we build our systems: we are privileging

Figure 2. Cindy Cummings in half/angel's *Spinstren*. © half/angel 2001.

certain kinds of information and discarding loads of the empirical data that are usually the core of these systems (things like xyz position, and so forth).

JGE: I stand in the wings clasping my peg top, watching the video of the field beside my house. As the image fades I walk into darkness. This piece, which began in a text that isn't here in performance, finds me standing with the top in my hand. I slowly look up. I begin to glance around me. I've lost a song, and as I try to hear it in the air, so I brave a tiny phrase, only to lose it again. After a while, Richard brings in the environment so that I perform a plural utterance—one live, the other mediated. An audience has difficulty distinguishing between the two kinds of voice, because they sound the same. But they know that there are too many voices for one body, even as the manner of their utterance is connected to my movements whether I actually voice them, or whether I 'move' them. Such voice ghosts perform loss, as I yearn to find the slipping song, only to sing with myself. Gradually I build up the interplay of live voice with pre-recorded sample, until I throw the peg top. The scarlet top turns in the air, lands, and spins impossibly. As it spins, so I find the song and shift into full-bodied and singular voice (Richard takes out my room of ghosts), a song that ends as the top slowly looses from its spinning, and topples.

Generating text

From the beginning, we had always felt a need to work with generated texts—words and poetic phrases cultured and spoken by the machines we use. Richard has often used (and continues to use) generated music within our work—there is something about the liveness of this that fits within the general sphere of the work; often tweaking and prodding happens during performance. Generated text involves the use of various software systems which analyse style, vocabulary and grammar, in order to generate 'original' texts. Such texts often have a strong taste of their original, but are also quite different. It is possible to alter the degree of autonomy the software has over composing its own texts. It is also possible to combine various texts. Our early experiments included combining Ovid's description of the myth of *Arachne and Athena*, with Grimm's version of *Sleeping Beauty*. The idea behind such experiments, was that we might also be able to weave with writing itself. As we become more proficient in these technologies, we begin to develop ways in which such texts might be generated in the moment of performance. For *Spinstren*, these experiments resulted in vertical projected texts, generated in performance—one of *Arachne and Athena* in Ovid's original Latin, another of Perault's *Sleeping Beauty*, and a third of one of the Carla texts.

The early research dealt with attempting to generate text in real-time. Actual generation of text, using Markov techniques to create new texts from known language/style bases was relatively trivial. What we wanted to do, however, was to generate spoken text in realtime—spoken, performed text is, after all, so much the basis of our work. This meant providing a grammar by recording a segment of text, cutting it up into word fragments, and making those fragments available as a library. We attempted to generate language information (i.e. grammar) from these fragments, but of course their original syntax was lost once they were broken up into individual words—we could only generate word frequency counts rather than actual grammar. This worked—technically—but Richard was very unhappy with the result. The text sounded entirely artificial, simply because the intonations of the individual words, taken out of context, were artificial. Platform information in many railway stations is now generated remotely (literally coming from some other part of the country) by stringing together samples of individual words. They are carefully intoned (performed) so that they are emotion-neutral, and even largely intonation (pitch)-neutral. Speech patterns are well-designed and constructed in such a way that it becomes possible to patch words together in any order and have them sound somewhat natural. Poetic language, however, doesn't conform well to this kind of processing. It became apparent quite quickly that attempting to generate a spoken poetic text, and keep a sense of the poetic in its performance was not easily possible. We decided to abandon this approach and find other ways to work with text generation.

In 'Geometries' from *Spinstren*, two performers share a text which is broken up with repetition. Such verbal fracturing also connects to the movement of the other performer. This is not generated text, of course, but it is nevertheless a narrative re-working of the text such that apparently random repetition and re-iteration of words and phrases are used, moving in the tide of the narrative's direction.

Richard also made a series of generated texts on video that were to be used in performance. Once again, this is not being generated by the computer in the space of performance; the text we see is the result of a process in which Richard made a file from each word of a text, designed as a vertical stripe. Each word lasts only 5 frames (one fifth of a second), except for what he chose as 'keywords'—words he considered important to the narrative. Some of these words are elongated to one second (although they instantly begin to fade once they appear on the screen, so they are not actually visible for the entire second). In practice, this created a blurred, illegible series of words flashing by on the screen, with some of the keywords being legible or close to legible. Because they are vertical they are even harder to grasp (and switching them to the horizontal made them too concrete, too easily narrative, too real). These columns of words construct a narrative of falling, of being just beyond one's grasp, even as their glimpsed phrases tell another counter-narrative. Like the teasing of wool to open its fibres, this process produces unusual archaeologies of familiar and less familiar texts, to produce tumbling phrases, and odd rhythms of stasis and percussion that are always different. Perhaps this isn't so much as storyweaving, as storyspinning.

Generating song

RP: Almost all the music heard in *Spinstren* is made in the moment, based on algorithmic recipes that twist and turn tonalities and tunings; that bend the melodic line; that make phrase lengths that are quite unpredictable in their nature and shape. Ironically, the pre-cogniscent machine is left to make decisions that ought to be solely in the realm of the composer. Perhaps it's a certain lack of alpha male genes, but I have long ago given away the composer's secrets, and sold the power to dominate. In building performance systems that are live and (sometimes) interactive, I'm giving away any final say on what the soundworlds will be. (This is, crudely, a feminised composition space. When others ask me if I'm uncomfortable with the gendered space of much of our work, I can quite easily say no.) In *Spinstren*, when a performer is not controlling and making the soundscape, the computer is making the final decisions not just about what notes we hear when, but where they are placed in the space. Of course, I haven't given everything away. I still have control over the ingredients of this recipe—I can still control the inherent shape of the sounds, how they will spin around the space, how their tunings will be bent and shaped. I make tables of melodies from which the machine-tunes are derived; I absolutely control the quality of the sound; I keep open ends and final bits of control that I manipulate in the performance. Just as in the interactive sections where I can control how the system is reacting, so too can I control the overall qualities of the melodies that are being shaped; where and how they place themselves into the space. But the control is not absolute, as such absolutism is not part of the spaces we create.

*

'Halving angels' serves as a metaphor for a process of poesis that operates in our collaboration and finished work. Halving angels is a process of examining what flies, falls, and is stolen in the poetics of making work. This is a book about technology. For us, technology is part of the poem, just as it makes a different poem possible.

People often ask us why our company is called *halflangel*. There is an old joke in which I reply 'well, I'm the angel, and Richard's the half.' But Richard is tired of this joke. Perhaps a different version might cheer him up: 'Why are you called *halflangel*?' 'Well, Richard and I are both quarters, and the angel is technology and poetry entwined in a magnificent heavenly embrace!'

*

I am falling slowly. I am in the middle of falling and able to think inside my own troubled descent. I am unprincipled in this falling. It is a kind of downwards, outwards endless stumble. This happens to me. I don't choose to fall. Falling approaches, and suddenly I am inside it.
Am I fallen?

Notes

1 *Spinstren* was funded by the Regional Arts Lottery Programme/South West Arts, the Arts Council of England, the Arts Council/An Chomhaírle Ealaíon, the Esmée Fairbairn Foundation, and the Arts Faculty Research Fund / University College Cork.

2 We often make our pieces by having different sections written on postcards stuck to a wall, and then we move them around to experiment with different compositional possibilities. Towards the end of this process, things that haven't found their place will often get cut, or thrown on the floor.

3 This is Cindy Cummings and Mary Nunan, both independent choreographers and performers in *The Secret Project*.

4 This connection is taken from Cixous' extraordinary 1975 text *The Laugh of the Medusa* in which she elaborates the compounded meanings of the French word 'voler,' which means both 'to fly' and 'to steal'. JGE first read this text as a young graduate student in 1988. Its wide passionate geography has continued to inform her practice, in particular around the notions of flight and theft as they relate to femininity:

> *Flying is a woman's gesture—flying in language and making it fly. We have all learnt the art of flying and its numerous techniques; for centuries we've been able to possess anything only by flying; we've lived in flight, stealing away, finding, when desired, narrow passageways, hidden crossovers. It's no accident that voler has a double meaning, that it plays on each of them and thus throws off the agents of sense. It's no accident: women take after birds*

> *and robbers just as robbers take after women and birds. They go by, they fly the coop, take pleasure in jumbling the order of space, in disorienting it, in changing around the furniture, dislocating things and values, breaking them all up, emptying structures, and turning propriety upside down. (Cixous 1975 258).*

5   Perhaps this is also a cultural preoccupation with angels; in the UK there are at least three other performance / arts admin. companies with 'angel' in their name; *Artangel*, *Third Angel*, and *Company of Angels*.

6   In an (unpublished) interview with Richard Povall, September 2000.

7   What I mean by this is that the pre-recorded vocal sample belongs to the past, and the live (miked) utterance belongs to the present. In this kind of performance, these two temporal spaces are confused intentionally. This is possible because they sound qualitatively the same (both are amplified), and because the physical gestures of utterance, are mixed up with other kinds of choreography, which themselves trigger pre-recorded samples of utterance. This is a confusion between voicing and moving. The programming of these intelligent environments is designed so that the computer can 'see' subtle as well as larger scale gestures, according to the 'taste' of the piece.

8   Our editors tell us this sentence is 'rather opaque', but I rather like it. When I use the phrase 'technology's digital duplicity' I'm referring to the binary nature of the digital, by playing on the term 'duplicity' meaning both 'double' (like a binary structure) and 'shady' or 'double-crossing'. I like this playfulness because our use of such technology is both through a process of programming numbers in a digital framework, and by corrupting such apparent order with a poetic gesture. '(T)he trouble of poetry's metonym' refers to metonymy within poetry, which is the process of referring to a part of something in order to suggest its entirety. Metonymic structures are horizontal, like the ripples of a pebble entering water. They are, by definition, unfinishable. This is in contrast to the vertical structure of metaphor, where one thing clearly stands for another; this thing replaces that thing. Metonymic structures allow a reader / audience to enter into a work differently than in metaphoric structures. I believe that our work is powerfully constituted by metonymic structures in language, sound and performance: I will give you a small part of something, but I will not tell you what that thing is, and it is that that will continue to unsettle you. This is a very particular kind of invitation into the work. '(P)oetry's metonym' is 'trouble' because of this unfinishablity, it can go on and on resonating in you, because we only offered a fleeting part of it, and so you are left wondering what it was, or even if you saw it at all. And finally, the "double trick" is the collaboration between Richard and I, which produces such metonymic structures (in part) from a binary machine called a computer.

9   This is one major limitation of using video-based systems, and one that we aim to look at in future research. We are hoping that using an infrared system may solve visible light dependency issues.

References

Ascott, R. (ed.) (1999) *Reframing consciousness.* Intellect ,Exeter, UK.
Cixous, H. (1975) The laugh of the Medusa. In Marks, E. and de Courtivron, I. (eds.) *New French feminisms.* Harvester, Brighton, pp. 245–64.
Merleau-Ponty, M. (1964) *The primacy of perception.* Northwestern University Press, Seattle.
Winterson, J. (1987) *Sexing the cherry.* Vintage, London.

7 The Performance Animation Toolbox: developing tools and methods through artistic research

Jørgen Callesen, Marika Kajo and Katrine Nilsen

Introduction

The Performance Animation Toolbox project is an artistic practice-based research project carried out in the Narrativity Studio at the Interactive Institute in Malmö, Sweden. The project was based on a series of performance animation workshops held from autumn 2001 until summer 2002, followed by two experimental productions in autumn 2002. The core research team was two researchers from the Interactive Institute, Marika Kajo and Jorgen Callesen, and stage designer Katrine Nilsen.

The aim of the project was to create a 'digital theatrical art laboratory' to investigate and develop the artistic potential of theoretical concepts within digital dramaturgy that involve real-time animation, tracking systems, role play, virtual puppets/actors and responsive set design. We wanted to make such techniques both functionally and financially more accessible, and to make artists realise how to implement them in their work. The long term goal is to integrate performance animation into modern dance, physical theatre and puppet theatre starting from the performer's traditions, abilities and knowledge.

The project drew from our experience with the theory and practice of improvisational theatre, pedagogical drama and puppet theatre to ensure that the technological concepts were grounded in a theatrical tradition. Through collaboration with stage designer Katrine Nilsen and a range of performing and visual artists, we wanted to develop new types of theatrical situations that integrate the possibilities and aesthetics evolving from digital media.

The main challenges were to bridge the gaps between research and production, and between theory and practice, involving professionals on all levels. It should not be theoretical studies supported by simple prototypes and it should not be full blown productions with very limited opportunities for experimentation—but something in between.

The project went through the following steps:

1. the establishment of performance animation as a model for interdisciplinary work;

2. the development of a technical set up, and a combined design and research method;
3. the definition of theatrical communication within mixed reality environments;
4. experimental research based on practical workshops in collaboration with artists;
5. the production of two experimental performance pieces.

Performance animation as a model for interdisciplinary work

The project evolved from our individual theoretical and artistic approaches and we first had to test some of the ideas through the development of possible theatrical situations. Through a series of meetings with artists, designers, technicians and platform developers from the Interactive Institute we created a method and a modular toolbox concept based on performance animation technology.

Performance animation is a term adapted from animation film production, where motion capture data recorded from a performer's movements is used to create 3D-animation for film, computer games, TV and advertisements. In this project the term is used in a much broader sense, covering various aspects of physical interaction influencing real-time computer generated sound, film and animation projected onto the stage.

So far this field of artistic practice is new territory where the borders between technology, dramaturgy and aesthetics are blurred and highly dependent on each other. To us it is a world of unexplored artistic possibilities, new concepts for physical interaction and theatrical communication, but the field is still very immature in relation to stage arts. To gain an overview we started by making a map

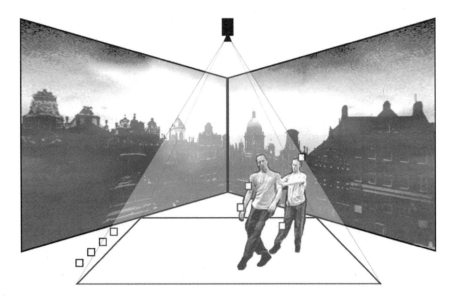

Figure 1. A responsive set design influenced by the performers movement captured through video tracking, body sensors and sensors on the floor (see top view in Figure 2).

of the interdisciplinary field between electronic engineering, theoretical research and artistic practice based on a simple theatrical situation, where a performer is influencing the set design through sensors (Figure 1).

Every running interactive system can be broken down to the three categories: sensors, dynamic systems and audio-visual expression (Table 1). We wanted to give the artists involved a chance to consciously know 'what happens inside the black-box' and the consequences for their dramaturgy and aesthetics. Our model enables the whole research team to get an overview and to develop a mutual understanding of the general principles, separate from any chosen technology.

	dramaturgy		*aesthetics/design*
t e c h n o l o g y	*sensors/mapping techniques*	*dynamic systems/databases*	*audio/visual expression*
	Magnetic MoCap	Inverse kinematics	2D/3D graphics
	Video tracking	Steering behaviours	Character design
	Infra red	Games / AI	Digital set design
	Joystick	State machines	Trick film
	Temperature	Self organising maps	Sound design
	Touch pads		Electronic music

Table 1. Performance animation as an art form combining technology, dramaturgy, aesthetics and design.

Developing a technical set up and a combined design and research method

Most of the tools for performance animation on the market are developed for professional character animation in film and computer games. It is new to use the techniques in a live situation where the performers are directly confronted with an audience or where the audience are invited on stage to use the equipment.

At present, performance animation technology is very complex and fragile. Compared to traditional theatre technology, it requires high-level specialists to operate . In many cases the tools cannot be implemented for the stage using 'normal' production methods. The creative team therefore often has to build their own tools.

In some cases standardised tools and media formats are too specialised for the needs of the stage, and there is then a need to reconfigure standard software for one's own purpose. In other cases the platforms are very open and contain possibilities that may not have been tried out by the software designers themselves, and there is a need to find a particular configuration and optimise it to a professional standard.

We wanted to create a method for a larger group of artists to develop and share their own customised tools, prototypes and experience in a coordinated mutual process. In this way the technology can be developed and tested from dramaturgic and aesthetic principles in a dialogue with electronic engineers and software developers.

The technical set-up

Setting up a format for theatrical experiments is like building a new type of stage, with new possibilities and restrictions. The basic elements were: real-time animation (1600 x 600 pixels) creating a back-projected image 7 x 3 metres on a screen; a floor space of 3 x 4 metres; stage lighting for the video tracker; and stereo sound (Figure 2).

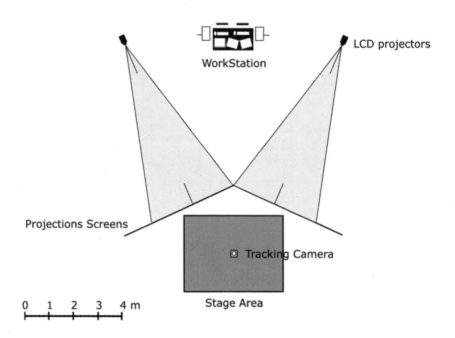

Figure 2. The technical setup defining the stage in the Art Lab (top view of Figure 1).

The audio-visual engine was based on the multimedia platform *Macromedia Director* with the ability to deliver real time animation and montage of a large variety of digital media formats controlled by a high level scripting language (*Lingo*). The choice of design was partly influenced by the limitation of the technical setup and the choice of *Director* as a real time animation engine.

The virtual design and the video tracking system

We decided to work with simple dramaturgic structures and ended up developing a number of prototypes for responsive set design and characters. In the design the style was a kind of 'virtual cardboard' design, very similar to old-fashioned puppet theatre, with different layers of two-dimensional set pieces and mobile elements or characters. This made it possible for us to create sceneries, with prefixed animations and different interactive elements.

A video tracking system called *Basker Vision* (programmed by Johan Torstensson) was developed to support interactive theatrical situations. The tracking system feeds information about the performers on stage to the animation engine which generates a response. The video tracking system was designed to track colour codes worn by the actors such as *Basker* hats, full body dress, objects, etc. Their identity, position and relation to fellow actors and space were identified through the measurement of distance and view angle. The system handled three actors, the minimal size of a group, and was chosen to limit the complexity of possible interpretations of relations and movement. This also kept us within the technical limitations of bandwidth and CPU, so we could achieve acceptable live response for the actors on stage (Figure 3). All in all this was a very simple low-cost and low-tech set-up, but it proved to be fast, flexible and a good prototyping tool for the development of new staging techniques, tracking principles and media formats.

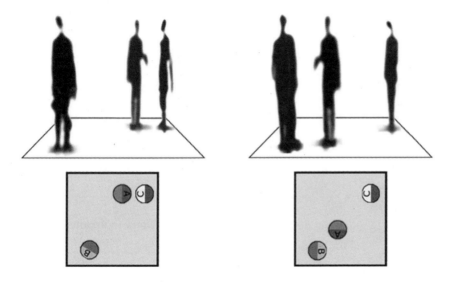

Figure 3. *Basker Vision*: a video tracking system registering the relations between the participants on stage.

Developing the research method

During the project we undertook a lot of activities, some in parallel and some in succession, ranging from free artistic experimentation, tool building and creative production, to theoretical discussions, methodology discussions and evaluation. For this there are no set methods and experiences to draw from, which meant that we had to define the methods as we went along, based on the backgrounds of the participants.

In the last evaluation session we created a more formalised process model giving an overview of the different phases in the project, showing the different hybrids between research and production, theory and practice (Figure 4). These hybrids could be a starting point for the development of a method for artistic practice based research.

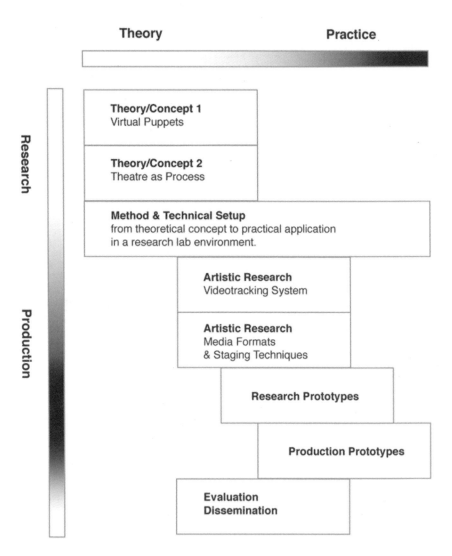

Theory   **Practice**

Research

Production

Theory/Concept 1
Virtual Puppets

Theory/Concept 2
Theatre as Process

Method & Technical Setup
from theoretical concept to practical application
in a research lab environment.

Artistic Research
Videotracking System

Artistic Research
Media Formats
& Staging Techniques

Research Prototypes

Production Prototypes

**Evaluation
Dissemination**

Figure 4. A process model for practice based research.

Theatrical communication in mixed reality environments

When you establish a virtual dimension on stage you deal with complex relations between virtual and actual representations of characters, space and the audience. The greatest dramaturgic challenge is how to make these relations clear and use them consciously when staging a play or a performance for an audience.

    With the points of departure in our individual backgrounds, we took three conventional theatrical approaches to the experiments.

1. Puppet theatre, which deals with the materiality of the medium and different levels of representation.

2. Improvisational drama and pedagogical drama, which deals with the involvement and activation of an audience.
3. The concept of 'the impersonated stage', which deals with anthropomorphic space, blurring the border between space and character.

Following Brenda Laurel (Laurel 1993), who used Aristotle to develop a new understanding of the computer medium, we found it useful to base our research on an established description of the minimum theatrical situation: A (the actor) plays B (the character) for C (the audience) (Bentley 1976).

This model for theatrical communication is not introduced to try to define new forms of theatre, which would go far beyond Bentley's conception of the theatre. Rather, it is used in the experimental process as a method to maintain focus on the theatrical elements in the experiments. We see this simplification as a way to combine the three different theatrical approaches and relate them to the formal rules of the tracking system and the computer-controlled characters and space.

The concept of the virtual puppet

The concept of the virtual puppet is developed in the performance animation project *The Family Factory*, based on the theory and practice of the contemporary puppet theatre (Callesen 2003a). In this context, the term 'puppet' is understood as an object which is used to communicate something to somebody by representing something other than what it is. In puppetry it is well known that the audience's ascribing of life to an object does not depend on how well the object portrays an original. Neither does it depend alone on the obvious factor of the skilled use of puppeteering technique. The crucial point is the way in which the player relates to the object, not only formally and technically, but socially and emotionally as well. From a dramaturgical point of view another level of representation can emerge from these relations, the story told being not only 'the story of the puppet' but also 'the story of the puppeteer'. Puppet theatre is in its core a theatre of more than one level of narration.

The other important factor about puppetry is the materiality of the object: form, colour, weight, surface, smell, texture and history. The expression of an object on stage results from a synergy of all these factors. Social and emotional experiences, common to both the player on the stage and the audience, creates a base to recognise one's own experiences and invest one's emotions. In this way the audience can ascribe the puppet (human) characteristics and personality and see the puppet as a character with independent will, opinions, intentions and emotions (Lund 2003).

The concept of the virtual puppet suggests that virtual characters in the projections are also puppets made out of a special material, which is light on a screen (Callesen 2003b).

In *The Family Factory* four virtual puppets operated by puppeteers were projected onto stage in a performance together with human actors. The puppeteers were professionally trained to control the puppets and made them seem 'alive' by making them relate to their surroundings in different ways. This made the communication between the human characters and the virtual characters convincing and understandable to the audience.

It is possible to create computer generated figures which are very lifelike models of living beings. Researchers in autonomous computer-generated behaviour are developing techniques to make virtual characters not only seem lifelike, but also seem rich in personality. The concept of the virtual puppet suggests this to be only one part of the problem. If the virtual puppet is controlled by a computer, the computer program will not only have to move the puppet in a lifelike manner, but will also have to do what the puppeteer does to the puppet: make the puppet behave and react to allow the participant to relate to the puppet socially and emotionally. These relations will have to be defined and used consciously.

The theoretical assumptions and the practical experience from *The Family Factory* project indicates that the material level of the virtual representation and the possible relations between objects, characters, actors and space (virtual or physical) is the foundation on which the possible dramaturgical constructions can be made. By using the puppet theatre as a theoretical point of departure it is possible to describe the different levels of representation in the virtual medium and the use of autonomous behaviour in a dramaturgical perspective (Andersen and Callesen 2001).

Theatre as process

The project also tried to integrate the audience on stage by working with theatre as a process. Bringing the audience on stage establishes a difficult theatrical situation. An actor (A) is trained to play a character (B), whereas the participating audience needs help to understand whether they are represented as themselves (A) or playing a character (B) for someone (C). Elements of the dramatic environment have to inform them of what to be and what to do, when to act and when to watch. The group also has a mutual responsibility to inform each other and to create the dramatic fiction together, as you do in improvisational theatre and pedagogical drama (Spolin 1963). The audience is put in a double role since they have to participate in the creation of the theatrical situation and to observe and experience it at the same time. In this experimental framework we call the new audience 'participants', but this is not satisfactory. In the future we need new terms to define how the participant relates to the complexity of relations and levels of fiction when experiencing theatre in mixed reality environments.

In an interactive drama, as well as in drama pedagogy, the intention and dramatic rules are normally introduced and skilfully managed by a director. We wanted to investigate whether a virtual actor (V-actor) can take over parts of this role and function as an artificial coach, represented by a virtual character or a responsive environment/scenery.

This V-actor has in many ways the same characteristics as the virtual puppet in the sense that the computer can take the role of the puppeteer. To become a proper V-actor it should have autonomous behaviour enabling it to possess the openness and ability to give and take or to imitate on the same conditions as the physical participants and thereby participate in the play and improvisation.

The final staging takes place as a process between the formulated dramaturgy and the participant's choices and intentions, in a mutual give and take. This principle is known as 'devising theatre'—good exponents of it are the British group *Forced Entertainment*, and it was used as a principle for interactive performance in the digital theatre project at Århus University in 2001 (Kjølner 2002, Etchells 1999).

The integration of a virtual coach in a performance demands certain capabilities of the participants and the design of the V-actor. To give a sense of a 'live meeting' between the V-actor and a group of participants on stage we wanted to try out different kinds of autonomous computer-generated behaviours for the V-actor. Through the use of different technologies, such as self organising maps (Kohonen 1997), games, artificial intelligence and steering behaviours (Reynolds 1999) this might become possible. This is a long-term process where the technologies have to be integrated into staging techniques and developed through practice, trying out the possibilities step by step.

The main focus in this project was on the development of a video tracking system, analysing group behaviours, and on different artistic approaches to the embodiment of the V-actor.

The impersonated stage

For the stage design we developed the concept of 'the impersonated stage' to refer to the consequences and possibilities of the new technology and the ideas of interactive staging. The challenge was first of all to impersonate and transform the universe of a drama into a visually and emotionally interesting and functional medium for physical interaction on the stage. In 3D computer game engines it is possible to give the illusion that you are surrounded by space, because you are fixed in front of the screen. But how do you embody and translate this into a physical space without simply creating empty landscapes beyond the walls? The task is to establish a design with some kind of presence and situations with direct confrontations that encourage playing and acting.

The first important step was to clarify the dramaturgic settings and mechanisms and to make them work. At first the possibilities of our set-up seemed very restricting compared to those that used high-end equipment such as magnetic motion capture and real time 3D animation, but it soon became obvious that the challenge does not lie in the visual finish or the performance of the technology, but elsewhere .

When we started the experiments in the set-up we therefore built a simple and clear structure, almost a kind of mechanical 'engine', as we wanted it to be as physical and emotionally logical as possible to make it easier for the participants to understand and learn how to control the virtual set and characters. Through the involvement of participants and V-actors on stage, the designer could relate directly to the new theatrical situation. Even though we had a long term ambition to stage a play, we agreed that the new relations had to be defined and investigated before the actual character development and design process could begin. In this way the designer had to postpone her ambitions to create an elaborate visual design and instead partake in the development of the technical and dramatic setup. By testing different prototypes with very simple themes and visual embodiments, the technical and dramaturgic principles became clear to us, after which we could start optimising the design and increase the complexity.

Experimental research based on practical workshops in collaboration with artists

In the first workshops we went through a series of very free and experimental sessions trying out a lot of audio-visual staging and playing techniques, developing a repertoire to draw from in the workshops that followed.

CityWalk

In *CityWalk* (Figure 5) we used some ideas from handling backgrounds in computer games and converted them to something that we thought would work on stage. The material for the animation was a scanned photo of the Copenhagen skyline showing rooftops and chimneys. The scenery was created as two 'virtual cardboards', a sky and a the city placed one in front of the other, like set elements, to give the illusion of depth and perspective.

This was hooked up to the video tracker and animated by the movement of the participant in the following way. If you stayed in the centre of stage nothing happened, you were at rest, but as soon as you moved away from the centre you would start to move around in the city. This involved moving a photo, twice as long as the visible scenery, in the opposite direction as the performer. The further you moved away from the centre the faster you would move over the city. This seemed to be a very efficient format for a responsive digital set.

Furthermore we wanted it to be possible to walk or zoom into the city. But this did not work out very well in this particular technical set-up, so instead we made it scroll up and down, so that the whole town seemed to rise when you approached and sink when you moved away.

CityWalk was effective because the participant gets an immediate response by moving around the space and therefore easily learns how to play with the effect. The dark and gothic style gave it weight and substance. It was suggestive, imaginative and created a sense of presence. We concluded that it had some dramaturgical potential: it could be useful both for a group of participants, and for a sole performer with the cityscape as a responsive backdrop.

GroupSound

With reference to devised theatre the aim with this prototype was to embody a V-actor as a device for interplay with the group of participants. We set out to embody the V-actor as a visual responsive impersonated space that would change dynamically depending on the behaviour of the group on stage (Figure 6). We soon realised there is a problem in balancing attention between the group and a projection. In many cases the images in the projection would attract the attention of the group, rather than stimulating the participants to relate to each other. By creating a non-visual soundscape we solved this problem because it made possible ambient and imaginative omnipresent expressions of space.

In the first prototype of *GroupSound* the space was divided into zones with different sounds. The zones were circular starting in the centre of the stage and expanding outwards. The quality of sound changed from a positive to a threatening sound as a person moved outwards. If a participant was present in a zone its particular sound would be activated. The number of actors within the same zone determined the volume of its sound.

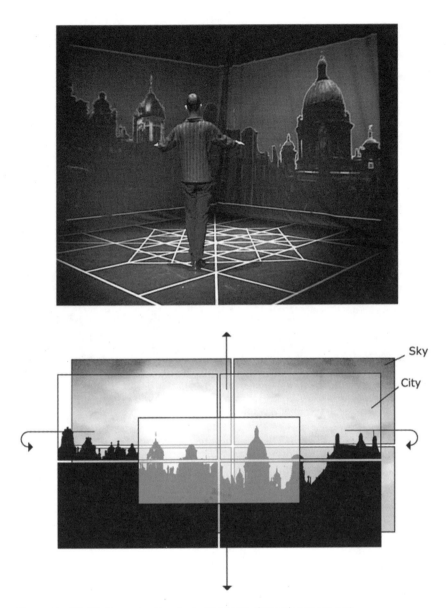

Figure 5. *CityWalk*: navigating in a cityscape through movement.

In tests the soundscape proved to be a good way to direct the focus towards play and negotiation among the group, stimulating varying improvisations. Gradually the participants developed an intuitive feeling of how to manipulate the soundscape, both as individuals and as a group.

Red sculpture
The use of physical objects as an interface is another way of directing the focus of the group and working with physical material as a link to a virtual soundscape. This was tried out in a performance art project in collaboration with sculptor Pierre

Figure 6. GroupSound: a responsive soundscape for group improvisation. The visual representation shown in the picture illustrates the principle used during the construction. In the finished soundscape, the screens are dark.

Treille and puppeteer Gabriel Hermand-Priquet and the *KeyStroke* software team at Ecole supérieure de l'image (ESI), Angoulême, France in 2002. This project used video tracking and a multimedia platform *KeyStroke* that worked through similar principles as our setup for the *GroupSound* prototype. Instead of tracking the participants on the stage we tracked a flexible red sculpture manipulated by two performers (Figure 7). Different positions and configurations of the sculpture would then trigger and modulate a digital soundscape, designed for the piece, which was performed for an invited audience.

The following description is taken from the evaluation report of the workshop.

> The austerely choreographed, semi-ritualistic duo of the two dancer-manipulators (Jørgen Callesen, Gabriel Hermand-Priquet) during the Friday evening performance generated an effective embodied counterpoint to the interactive sonic space, which evolved with the movements of the red sculpture. Simplicity of the movement sequences and sound components allowed the audience to quickly and effectively grasp the logic behind the various elements (performers, sculpture, sound), enhancing engagement in this interactive dance performance.
> (Norman 2002)

This experiment showed that an audience is able to grasp quite complicated abstract relations between physical and virtual levels of representation if they are choreographed and staged using known conventions—in this case a hybrid between sculpture, dance and puppet theatre.

Figure 7 Red sculpture: performance art in a sonic space.

Research prototypes

The problem with artistic practice-based research in academic environments is that the research team often lack the practical knowledge, experience and artistic skills to be able to define the research questions and carry out meaningful experiments, seen from the perspective of the production environments. In this project we aimed rather to find out how a person with artistic skills would define the research questions and carry out experiments in the set-up.

This was mainly done in a 3-day workshop with performers, technicians, directors and designers from Sweden's free theatre groups as part of the theatre festival KONKRET in Malmö, 3–5 May 2002. For this workshop we prepared a series of artistic prototypes to demonstrate the principle of performance animation in the particular setup. It was then the task of the invited artists to define a project and a research question in collaboration with the researchers. During the three days of the workshop we produced five different prototypes. The results were shown to an audience and discussed among the group with invited evaluators who had a background in theatre and performance. The results were finally published on a CD-ROM (Callesen, Kajo and Hallborg 2002).

Dancing Shadows

Dancing Shadows was made as a collaboration between modern dancer Peter Jul Nielsen from Copenhagen and Jorgen Callesen. It involved a trick film, showing a pre-recorded sequence of the dancer, which was animated by two principles: the speed in which the trick film was replayed would follow the speed of the performer; and the position in which the film would be shown would move in the opposite direction as the performer (Figure 8).

The movements and speed of the trick film were influenced by a dynamic model, giving it elastic resistance, which made it fluent and added plasticity to the movement. *Dancing Shadows* was extraordinary effective in spite of its simplicity. By entering the space you got an immediate response and feeling of 'contact' with the shadow. It shows both the potential for a performance where dancers use the virtual shadow as a partner for a choreography and also for an installation that invites the audience to participate by stimulating them to move in space and even dance themselves!

The *SheMonster*

As in *Dancing Shadows* the principle behind this prototype was also very simple. When you walked across the room a woman would flirt with you but when you went closer to her she would gradually transform into a monster. This was done as a trick film consisting of 72 drawings made by animator Anna Kellert from the Department of Animation and Animated Film, Eksjo, Sweden. The playback of the animation was influenced by a dynamic model, to make it more fluent and to give it life (Figure 9).

This prototype shows that traditional frame animation can work very well for performance animation if it is designed for the space. It can be used by an actor or a puppeteer for a performance with a V-actor but will, of course, need to be developed and integrated into a story. It will also work very well with the general audience as participants, because even though the story is revealed very quickly it does not unfold its nuances until you start moving in space.

Heaven and Hell

With *Heaven and Hell* we moved from a responsive space to space as an abstract character with autonomous behaviour. The intentions were to create a mood space, where the space had an attitude to the group of actors and the way they behaved on stage. The dramaturgy was based on a gradual movement from positive to negative energy. The mood space should be controlled by the output from the *Basker Vision* video tracker parsing information about behaviour from the individual and the group as a whole. The idea was that it should react differently if the participants face each other, turn away, are close together, move slowly or fast, etc. (Figure 10).

The V-actor was not intended to be an artificial actor with naturalistic human features, which is often the chosen embodiment in many projects dealing with autonomous agents. Depending on the chosen dramatic universe the V-actor can be open for play and negotiation in many different embodiments such as an audio-visual character, as abstract material, anthropomorphic space, a physical object, etc. To investigate possible relations between a group of participants and a V-actor, two types of embodiment were designed: a responsive space; and an abstract character with autonomous behaviour. For the responsive space we discussed using ambient expressions as texture, colour and light but, with *Director* as the animation engine, we ended up making an animated abstract space based on the principles from *CityWalk*.

To focus the attention of the participants, a mystical virtual character, a kind of big black spheroid with autonomous behaviour, was introduced. The spheroid was then staged inside the responsive space and could be pre-programmed for

Figure 8. Trick film of dance sequences animated with movement from dancer.

different movement patterns, depending on which relations to the performer or the group we wanted to respond to. Its dynamic movements were created by autonomous steering behaviours (Reynolds 1999). In the first version it floated back and forth in a predefined horizontal path, repulsed and activated by the movement of the participant on the stage. It looked as if the spheroid was gently pushed and pulled around the stage. The scenery was an animated background influenced by the movement of the spheroid. By scrolling up or down it would bring you either into Heaven or down to Hell, as a result of interaction with the spheroid.

When designing *Heaven and Hell* we did not have high expectations of what it could achieve. Partly because we had not yet tested the final version of the video tracker, which could register eye contact and more advanced group behaviours. We therefore did not really know what we were designing it for. The available techniques in *Director* were so crude that it was likely that it would quickly reveal its almost mechanical nature, making it difficult for a trained actor to work with and uninteresting for an audience to watch. But when we finally tried it out, we were surprised to realise that its simplicity was what made it so easy to interact and play with, and we had a magic moment together!

Two experimental performance pieces

In the final phase of the project we had the opportunity to produce two experimental pieces based on different elements from the prototypes: an altered version of *Heaven and Hell* as a responsive set design for a dance improvisation, and the interactive ghost story *Spirits on stage*.

Heaven and Hell for an open dance improvisation

We were invited to test *Heaven and Hell* in an open dance improvisation with a group of three dancers at Skaanes Dance Theatre, Malmö, Sweden. For this event we designed two production prototypes working with dynamic relations between the group and the black spheroid. The space would then show its attitude towards the group by changing between Heaven and Hell according to the constellation of the group.

In the first version for three dancers the spheroid was programmed to act as if afraid of dancer X, but attracted to dancers Y and Z. The space would appear happy (Heaven), if the spheroid was close to Y and Z and unhappy (Hell), if it was obstructed by X (Figure 11). The dancers wore costumes in red (X), dark blue (Y) and light blue (Z). The spheroid was repulsed by the red dancer (X), who could scare it away whenever it tried to get near the two blue dancers (Y and Z). When the two blue dancers were alone on the stage, the sphere would try to go to the closer one. This way the two dancers could either fight to get power over it or mutually manipulate it like a string puppet. Each of the dancers could also interact alone with the sphere, according to whether they repulsed it or attracted it.

In the second version for a solo performance the spheroid was interested in dancer X, but afraid of direct confrontation (Figure 12). A single dancer wore a two-coloured costume with a red front and a blue back. The sphere would follow the dancer whenever he turned his back, but rapidly move away as soon as he

Figure 9. The *SheMonster* was also created at the KONKRET festival from an idea by director Christina Evers, Lilla Teatern, Lund, Sweden.

Figure 10. *Heaven and Hell*: the mystical spheroid in relation to the responsive space.

Figure 11. The spheroid is afraid of dancer A, but attracted by dancers B and C.

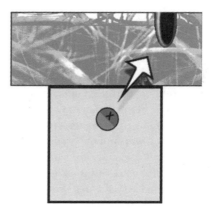

Figure 12. The spheroid is curious about dancer A, but afraid of direct confrontation.

turned and faced it. This made it possible for the dancer to create a game of flirtation and peek-a-boo by turning front or back. The dancers invented another variation of this, by introducing two more dancers, one wearing a red and one a blue costume, making it in to a rare kind of *pas de deux* or should one say *ménage à troi.*

In the performance we again experienced the problem of balancing attention between the group of performers and the projection. The V-actor constantly demanded attention from the performers. To improvise freely they needed a longer rehearsal period to develop an intuitive relation to the virtual space and character. This would enable them to work with it even when it was out of sight. Another way of solving the problem would be by developing other ways of integrating the projection into the set design.

Spirits on stage
In spring 2002 the research group was contacted by a producer from the Malmö City Festival, who wanted to involve installation and performance pieces with digital media in the programme. The Cultural Arena is located in a circus tent in

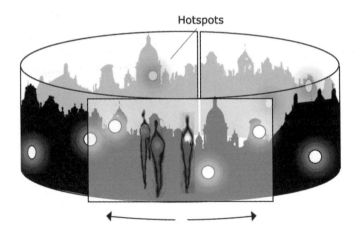

Figure 13. Ghosts, hotspots and 360 degree panorama.

central Malmö and seats 300 people. The repertoire is normally theatre, dance, rock music, stand-up and cultural events, but this year they wanted something new— something that involved interaction with the audience and related to digital culture and media. With a point of departure in our research prototypes, we decided to produce an interactive ghost story *Spirits on stage* for the circus tent. In this phase the method was far more production oriented. The goal was to create a great experience for an audience with the available material and tools.

Spirits on stage was designed as a narrative cityscape for audience participation as well as for a traditional passive audience. We agreed to make a prototype for an installation that stimulated audience members to perform on stage in a way that would be spectacular enough to engage the passive audience. We also made an agreement that the installation should be presented as an experiment since it would be the first time we tried out these staging techniques on an audience.

We decided on the *CityWalk* as a point of departure, introducing some kind of group dynamics in relation to a storyline. The installation would be on stage for only two hours, therefore the principles had to be simple and clear and the idea had to be easy to grasp to be sure not to lose the audience.

The result was a ghost story taking place in the city of Malmö, transformed into an enchanted town, that emerged at the stroke of midnight and vanished at dawn. Everybody knows ghost stories, they are small often mythical narratives related to certain locations and their history and can be linked together in a non-linear way. Furthermore the audience could easily relate to known locations from their own city.

The intention was to make a humorous, slapstick interpretation of everyday situations and places that had a special identity. The statue on the plaza would come alive, money would rain outside the casino and you would discover a corpse in the old bath house, see a flying U-boat in the harbour, an exploding TV in the council flat or dancing chillies in the shawarma bar. Locations were selected for their ability to tell a small story.

With the help of game designer and programmer Simon Løvind we optimised the *CityWalk* prototype to run smoothly with a 360-degree cityscape,

Figure 14.
(*Top*) The cityscape with hidden ghost stories.
(*Bottom*) A story unfolding: 'The water tower turns out to be an UFO'.

which was created as a photo-collage in four thematic sections with different locations and elements from around Malmö. Each section was identified by a significant ambient soundscape and held hotspots for 5 or 6 different ghost stories (Figure 13).

There was room for three participants at a time, who became their own ghost by dressing in a colour-coded 'ghost cape'. For each ghost cape there would be a ghost avatar in the virtual set. These ghost avatars were made out of light, following the person on stage like a torch. In this way each person was embedded in the story and was able to 'see' in the dark. A montage was made where only the contours of the city were visible. The ghosts would then light up the details by walking through the city, inviting the participant to explore the space and find the hidden ghost stories. To give each ghost an individual appearance and a character they were pre-animated in a short loop and marked by a distinguishing sound, activated whenever they touched one another. With the beating heart of the individual ghosts the three participants could find the active areas and trigger the ghost stories, which would only appear once during each session.

Each ghost story had three levels of presentation: a single sound or a slight visual change signalling the presence of a ghost story; a second sound level and an animated visual effect giving an introduction to the story; and the full story. No matter which ghost or collaborating pair discovered the hotspot, the first level appeared when a single ghost enters, the second level when two ghosts enter together and the third level when all three ghosts enter together.

For the participants to pan around in the cityscape a navigation control panel, designed as buttons for forward and rewind, was placed on the floor at the front of the stage; still within sight of the tracking camera, but outside the visible area of the virtual scenery. By standing on this control panel the participants were able to move the panorama to the left or to the right at different speeds. The different choices of navigation in the narrative cityscape gave interesting opportunities for the group to negotiate where and how to move together on stage. The participants' ability to play their part also had a great impact on the installation as playground as well as for the audience observing the event. Even though we ended up having a commentator introducing the participants one by one to their virtual ghost and showing them how to navigate, we noticed that after a while the rest of the audience soon learned the rules by watching from outside.

It was a different situation to be partaking on the stage than to be watching it from afar. By entering the stage the participants were part of the action, but they also lost the overview of the full picture because of the size of the details and the short distance to the projection. To regain the overview, they were forced to step back out of the scene, which brought them out of control with the action. This is an interesting dilemma of a participating audience which we would like to investigate further.

Conclusion

Having gone through all the phases in the process model, we have provided some theories, methods, tools and products that shed some light on what happens when visions about new theatrical concepts are tried out in real life.

In this project we did not produce actual works of art, but we have achieved something, which in our opinion is a necessary precondition. We have established a frame where theoretical, technological and artistic concepts can be tested under production-like circumstances.

On the question whether you can conduct this type of experimental research within the frame of normal production conditions or in academic research environments, we are likely to say, No. To develop the prototypes we have crossed a lot of boundaries between theoretical traditions, artistic disciplines and professional boundaries in production teams. This has taken a lot of resources, but has also given us insights in how to deal with different types of creative development in different stages in the process.

We have found it useful to distinguish between prototypes used in research and prototypes used in production and between different types of audiences for the prototypes. The research prototypes have the purpose to test the theoretical and technological concepts and thereby make it understandable for both artists and researchers. The actual running prototype and different ways to reproduce the experience when it is used in 'laboratory performances' can be seen as a vital and necessary part of a new publication and demonstration format. This is useful when artists and researchers need to describe, discuss and exchange their ideas and results, but is not to be mistaken by an actual performance piece.

The production prototypes are meant for a specific dramatic production where the technology and artwork are optimised to a standard, which can be tested on an 'unprepared' audience on real stages. Much too often research prototypes are presented to audiences from the public giving them an experience that can only unfold by understanding the research concept. We experienced this to some extent in the open repetition of *Heaven and Hell* at Skaanes Dance Theatre. After the repetition we realised that members of the audience had expected a full blown performance, even though it was announced as results from an experimental improvisation, followed by a presentation and a discussion with involved researchers and artists.

It is therefore crucial to understand that the final test of a production prototype lies in the perception of the audience. Just like in any other theatre or dance production it must be evaluated through a normal cultural process. In the case of *Spirits on Stage* we had a strong feeling that a majority of the varied festival audience, spanning from children, youngsters, families and a theatre goers, were amused and thrilled by the experience, which was also confirmed by the producer and representatives from the festival.

There are also scientific methods to deal with this question such as analysis of video recordings of audience behaviour and interviews with audience members. We believe it is problematic to use these methods to evaluate the artistic quality of a piece, but they can give very useful insight to long term research projects and evaluations of prototypes informing both artists and researchers.

The Performance Animation Toolbox is meant as a tool for planning, coordinating and building different prototypes in a way that they can be transferred from one phase to the next. It is also a way for artist and researchers to share experience, exchange results and benefit from each other's inventions, which is a necessity if the aim is to develop performance animation as an art form and transform visions into actual works of art.

References

Andersen, P. B. and Callesen, J. (2001) Agents as actors. In Qvortrup, L. (ed.) *Virtual interaction: interaction in virtual inhabited 3D worlds.* Springer-Verlag, London.

Bentley, E. (1976) *The theory of the modern stage: an introduction to modern theatre and drama.* Penguin Books, Harmondsworth, Middlesex.

Callesen, J., Kajo, M. and Hallborg, A. (2002) *KONKRET workshop; documentation and prototypes* (CD-ROM available from <joca@skydebanen.net>).

Callesen, J. (2003a) The family factory: developing new methods for live 3D animation. In Madsen, K.H. and Qvortrup, L. (eds.) *Virtual methodolgy: behind the scenes of multimedia production.* Springer-Verlag, London.

Callesen, J. (2003b) Virtual puppets in performance. *Proceedings of Marionette: Metaphysics, Mechanics, Modernity.* International symposium, 28 March–1 April 2001. Copenhagen University, Copenhagen.

Etchells, T. (1999) *Certain fragments: contemporary performance and Forced Entertainment.* Routledge, London.

Kajo, M. and Johansson, M. (2001), Common playground. *Proceedings of Cast01.* A presentation on the project Communicating Moods in Space.

Kjølner, T. and Szatkowski, (2002) Dramaturgy in building multimedia performances: devising and analysing. In Madsen, K.H. (ed.) *Virtual methodolgy: behind the scenes of virtual inhabited 3D worlds.* Springer-Verlag, London.

Kohonen, T. (1997) *Self organizing maps.* Second Edition. Springer Verlag, London.

Laurel, B. (1993) *Computers as theatre.* Addison-Wesley Publishing Co., Reading, Mass.

Lund, J. (2003) Puppets and computers. *Proceedings of Marionette: Metaphysics, Mechanics, Modernity.* International symposium, 28 March–1 April 2001. Copenhagen University, Copenhagen

Norman, S. J. (2002) ComiXlam Event Report/RADICAL. Deliverable D.4.2.1, Information Society Technologies Programme, European Commission.

Reynolds, C. W. (1999) Steering behaviors for autonomous characters. *Proceedings of Game Developers Conference,* San José.

Spolin, V. (1963) *Improvisation for the theater: a handbook of teaching and directing techniques.* North Western University Press, Illinois.

8 Theatre of context: digital's absurd role in dramatic literature

Jeff Burke and Jared Stein

Technology in theatre exists almost exclusively in the spectacle of performance, and though it has largely defined the societies of modern drama, it has rarely enabled new forms of dramatic literature. Technology creates spectacle, serves as subject, yet remains disconnected from the structure of popular and avant garde theatre alike. Directors and designers incorporate new tools and techniques to achieve certain effects in a piece's realisation, and newly constructed theatres are built to incorporate modern innovations that often prove useful and provocative to their artists. This trend has continued with the advent of digital technology, which enables amazing new uses of lighting, sound, and projection in performance. More adventurous directors and performers have experimented with emerging technologies enabling telepresence, 'virtual' theatrical worlds, and many forms of multimedia performance. The great body of this work consists of re-interpretation of traditional texts and creation of new works with either minimal text or none at all. It has not yet extended the rich tradition of dramatic literature, even the seemingly related epic theatre or the 'algorithmic' works of Beckett and others. Our own collaboration, between engineer and playwright, has led us to pursue this direction, explained below in the context of a particular work.

The *Iliad Project* is an ongoing research effort in the UCLA School of Theater, Film and Television to co-develop an original text, design, technology, and performance technique that focuses on the unique characteristics of digital technology as they affect dramatic literature and performance practice. After initial experimentation in design (Burke 2002), we focus on roles of digital technology that can be intrinsic to the script itself, provoking different interpretations by designers and directors rather than prescribing the entire performance experience. The text is extended to include algorithms defining relationships among 'inputs' to the play's world—images, sounds, demographics, and other information—from the audience's everyday world. This ability to include the audience as an important element of the story, and to do so via technology, has become a driving force of our collaboration. We see it as one pathway for the development of drama, making technology as intrinsic as Aristotle's logic and as captivating as Beckett's illogic.

Theatre artists have used emerging technologies in their reinterpretation of existing works—for example, David Saltz's *Beckett Space* at the State University of New York at Stony Brook (Saltz 1999) or Mark Reaney's work at the University of Kansas (Reaney 1999). The Wooster Group and Robert Lepage, among others, have staged highly sophisticated multimedia works with current commercially available technologies. Experimentation with emerging digital technologies in original works, however, has largely been the domain of performance art (Birringer 1999), which offers a rich environment for experimentation because of its holistic creative approach—spectacle, text, and the role of the audience are each designed to serve an overall experience. The vision of this final experience drives the development and integration of each component, including the script, allowing it to rapidly incorporate new forms of technologically enabled or motivated relationships between audience, performer, and author. It is well-suited to be a laboratory for exploring and challenging emerging technologies and their associated cultural perspectives.

Yet this focus on performance art overlooks dramatic literature's own fluidity—how it embeds the opportunity for reinterpretation by directors and designers. The disconnect between digitally-influenced performance practice and modern dramatic literature arises partially from a myopic definition of 'the digital' that focuses on a few particular input and output manifestations of that realm: projections, computer graphics, automated scenery, sound effects, and even particular types of sensing. Advances in each of these areas make media responsive and malleable in ways never before possible. But the power of 'the digital' lies in its broad scope and abstract nature—the fact that mathematical formulas control numerical processing to drive every element of spectacle mentioned above, not the particular manifestations themselves. 'The digital' is an abstract representational arena that can be manipulated at incredible speeds by man-made machines, enabling connections across modal and geographic boundaries, into huge stored datasets, and between anything that can be digitally represented. If it is appropriate for performance and installation art to simultaneously explore both digital processes and specific input/output manifestations, dramatic literature—specifically, the play—can incorporate the processes alone into structured text, leaving the particular 'input and output' open for reinterpretation by directors and designers.

The written play's construction for collective experience (a general feature of performance), its traditional use of mimetic narrative, of showing rather than telling, and its openness for reinterpretation by directors and designers suggest intriguing challenges for digital artists and engineers. For playwrights, emerging technologies allow autonomous systems to be implemented according to the rules of the play—embodying rather than illustrating its world. They can interconnect geographically separate elements and act in response to individual presence, identity, and action at a speed and scale that draws together many individual phenomena into a collective experience. Because digital processes operate on abstract models of the world, they can be designed and specified in a way that allows for meaningful 'reinterpretation' of their inputs and outputs.

Our first attempt at incorporating digitally enabled processes into dramatic literature, this *Iliad*, begins with a website specified within the script, one that replaces a more traditional prologue and is accessed by the audience before the performance from home or immediately outside of the venue. In addition to introducing the world of the play, it pulls information from the participants for the purposes of the characters. The minimum requirement for input is a direct survey of the web visitor, but more subtle information might also be measured using either custom-designed or commercially available software: What pages do the visitors look at repeatedly? What path do they follow? How do they read or experience this part of the text? The answers can be incorporated as directorial or design brushstrokes after the structure is established. A database must store the survey responses and usage tracking information for later use onstage, where the survey results, along with audience images and sound bites later culled and indexed by digital means are displayed and redelivered within the political rhetoric of the characters. The web site's setup and interaction continues via an e-mail news list, which also serves the characters' purposes. In relation to a more traditional dramatic structure, this serves as a transition from the prologue to the larger body of performance and includes an inciting incident: Helen's kidnapping.

Audiences arriving at the event's location encounter a gallery of photography that elaborates on the themes of the website, and includes more information relating it to the kidnapping. The gallery establishes their role as Achaean constituency and introduces a tame initial form of the Achaeans' recognisable manipulation of media. 'Photography' is the minimum requirement for output in this space; this detail is open to directorial reinterpretation that may incorporate digital technology, but no further spectacle is required by the play's action. The input requirement is that the participants' tickets allow them to be physically located (with very low precision) and identified in the gallery. This will enable the computer and sensor systems implementing the play's rules to keep records of each person's presence and action in the gallery alongside the existing information about their interaction with the website. A particular implementation might use Radio Frequency Identification (RFID) technology and equip the space with an image capture system to gather snapshots of facial reactions to the gallery's provocative photography for later use in the performance. This gallery functions as a play's more traditional exposition, but must also capture images and simple actions of the participants and store them for future use, gathering the first echoes of the audience's real lives into the machinery of the play itself.

By the time the recognisable 'performance' begins, the audience and the performers will be ready and eager to act upon the consequences of the inciting incident—the kidnapping—without the need for further exposition. The media of a given audience will begin to appear, recontextualised, within the performance: the performers will be supported and even directly prompted by the play's technological systems with information about the participants to dynamically relate to the given audience and situation. For example, when the politicians of the Achaean cause must refer to a map, the map derives its boundaries from the demographics of the attending audience, including their actual addresses. Though the map itself might take many forms in a final staging of the work, our realisation uses Geographic Information System (GIS) software to look up and create different images at the same point in performance for different audiences (see Figures 1 and 2).

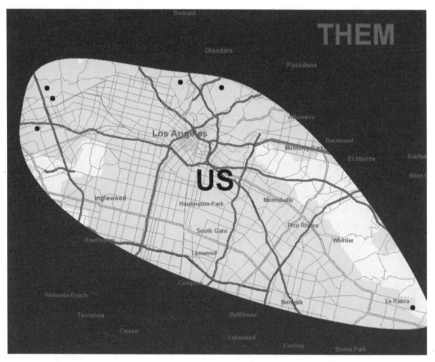

Figure 1. Dynamic rendering of a test map for one small audience.

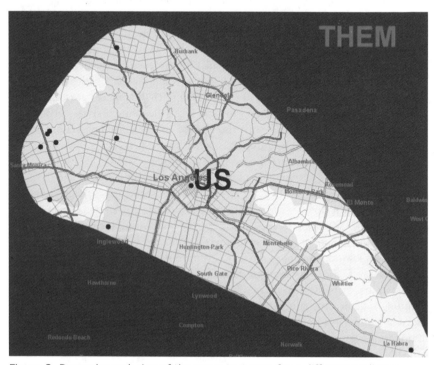

Figure 2. Dynamic rendering of the same test map for a different audience.

Marie-Laure Ryan characterises this type of relationship between viewer-participant and a work as 'internal-ontological': the participants are characters within the time and space of the fictional world. She goes on to state that "in this type of system, interactivity must be intense, since we live our lives by constantly engaging with the surrounding world." (Ryan 2002 601) The more demanding, direct interaction of the website and galleries is interwoven with the collective experience of performance, where the audience experiences the gradually building re-use and re-contextualisation of their information over the course of the play. The story built around the Achaean political machinery justifies each component, and oscillation between components slowly reveals the processes (or, perhaps, the 'poetics') of the piece to the audience, commenting through the story on the system of technology itself that constructs and manipulates the audience's experience using their images, actions, surroundings, and mere presence for the characters' and dramatists' purposes.

As the first traditional act concludes, an interactive intermission gallery—a carnivalesque abstraction of the battles at hand—sustains the event and, like the opening gallery, provides fuel for further dramatic development by observing and interacting with the audience. Technology again culls together the images, sounds, and statistics of an Achaean constituency (the audience) at play. As the second performance act begins, the piece turns on itself, and the audience finds its demographics used to the opposite effect within the story. The reversal is intensified by the acquisitions—echoes of the audience, live media from the city—gathered up until that point. Media from throughout the city of the performance is streamed into the space, and juxtaposed with recently captured images, processed website and gallery responses, and real-time sensor data. When the event concludes, a final gallery, customised to the attending audience, substitutes for a more traditional epilogue.

To describe such a performance, the dramatic text must encompass not only dialogue, situation and action, but define a system of digitally-enabled connections through which the dramatist draws relevance, either directly or indirectly, from the audience's environment and actions. Combining instructions for human action with algorithmic rules driving digital processes, the text becomes an architecture for performance in which both human action and digital processes intertwine the real and fictional worlds. The use of audience information and data's presence itself within the story is justified by the absurd logic of the characters and their world. Following and extending the progress of modern drama, the audience is integrated into the story in a way that makes them a point of relevance; their unpredictable, fleeting specifics are encompassed by the story but open up an immense variety of interpretations. A particular narrative provides the (hopefully) coherent and continual justification for the audience's technologically-mediated interaction with the play's world. The Brechtian idea that one has to understand the modern machinery of power before writing about it is carried a step further, by actually implementing an absurd but logical extension of the media-manipulation processes that create and maintain power today.

For 'interactive' digital art in general, a narrative approach offers an ancient and traditionally successful method of engaging an audience for extended interaction—now potentially in more than one space and at different levels of perceived control. Dramatic texts created with digital technology's capabilities for connection-making in mind can take this further, motivating the technology within the action as well as the design of the spectacle. Where performance art tends to decrease or remove text-based structure to gain fluidity of expression, here increased structure is used to justify technological choices within dramatic action, at some loss of fluidity. Justification within the narrative itself allows an audience to better consider the technology's implications within the story, without the distraction of an unclear motivation for its presence.

> *Digitality is a fluid environment; narrative, as a type of meaning, is a solid structure. To reconcile the two, some compromise will be necessary. Narrative will have to learn to share the spotlight with other types of sensory data… Conversely, the medium will have to give up some of its fluidity to allow narrative meaning to solidify in the mind of the reader.*
> *(Ryan 2002 607)*

Given this fluidity of the digital, why do we confine ourselves to only one narrative? In parallel with theatre and performance art's incorporation and criticism of technology, writers of modern literature have experimented with a variety of methods to digitally deliver non-linear text, commonly referred to as 'hypertext.' These two bodies of work have rarely intersected in physically staged works. Primary tools of hypertext authors are branching structures that enable non-linear navigation of a set of text fragments driven by user interaction. They allow multiple, varied readings of a text and find their precursor in literary works like Julio Cortazar's *Hopscotch* and Vladimir Nabokov's *Pale Fire*. This model is both supported and encouraged by the organisation of the world wide web, which allows text and media to be linked together in complex author-defined structures. It is also the underlying organisation of many single- and multi-player computer games now available. Our performance experiments with students suggest that performers delivery of dynamically assembled or chosen text (e.g. by hidden earpiece) is quite possible and interesting, so this seems a natural direction for exploration.

Dramatic literature's goal of specifying a collective experience confounds many techniques of hypertext, however. In hypertext and 'non-linear' novels like *Pale Fire* and *Hopscotch*, branching structures allow the works to encompass multiple points of view and provide many paths of discovery within the text. To some, this approach parallels the real world more directly than traditional forms: the rejection of a single plot "signifies the recognition that the world is a web of possibilities and that the work of art must reproduce this physiognomy" (Eco 1989 114). For audiences to experience this in practice as well as understand it in theory, however, requires them to actually explore multiple paths within a work—to 'reread' the text, following different paths and comparing the experiences. The concept of the user's increased power is present when multiple choices are given and one chosen. (Indeed, this will impact the reader's perception of the work even within a single reading.) However, the story's multifaceted nature is revealed only if sections are reread with different choices made.

Therefore, although digitally-supported voting on story choices—navigation of a dramatic hypertext—seems to offer the opportunity for 'collective' decision-making by the audience, those choices collapse structurally into a linear form when audience experience is considered, because there is no opportunity for re-reading.

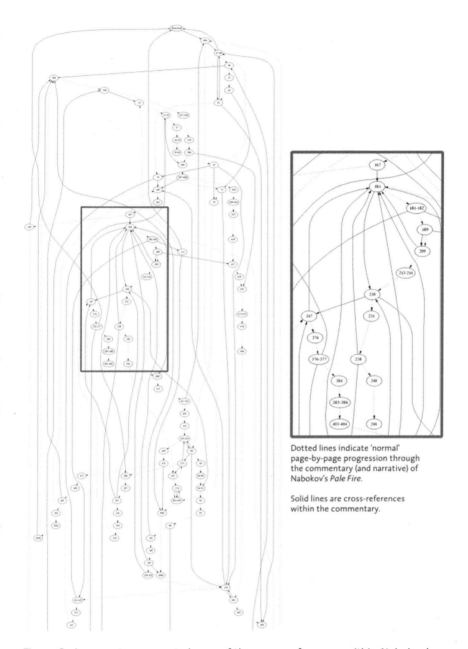

Dotted lines indicate 'normal' page-by-page progression through the commentary (and narrative) of Nabokov's *Pale Fire*.

Solid lines are cross-references within the commentary.

Figure 3. A computer-generated map of the cross-references within Nabokov's novel *Pale Fire*.

Figure 3 shows a map of the many cross-references between the sections of commentary, which is also the main narrative, in *Pale Fire*. Any single reading of the text, even one that goes back over certain sections, would trace only a single path. Though one could imagine a work in which actors re-performed certain sections in a different order, the self-directed discovery of the branching literary text does not appear to translate well to the stage. Performance renders a single 'reading' that is indistinguishable from an event in which no choice at all existed, except for the presence of the choice.

Unless its presence alone is necessary to the story, what purpose can voting on story choices serve without re-reading? What remains is the audience's presumption that their opinion might have mattered, the question of whether it really did, and the frustration of an often unjustified omnipotence. One of the authors recently had the opportunity to sit alone in an immersive multimedia experience—a cinematic-style educational game— designed for an audience of about forty people. Alone, his vote was all that mattered. Even in this case, in order to build dramatic tension and progress the story, none of the voting choices in the middle of this carefully constructed experience could solve the problem posed by the story, because if it had been solvable the story would have been finished (and incomplete) only a few minutes into the experience. Even if performers, designers and writers were enabled by digital technology to handle an unlimited set of choices, in order for collective decision to be feasible, only a few could be presented to allow consensus. In this case, how can the interactors "retain a reasonable freedom of action throughout the performance without taking the plot in a direction for which there is no ready-made, logically coherent response stored in the system?" (Ryan 2002 592)

The remaining motivation, then, is conceptual, and we at least are hard-pressed to engage an audience for any significant length of time through purely theoretical means. Like Ryan, we don't intend to

> ...*discourage conceptual art and experimental literature, but as long as...authors limit themselves to this route, they should not be surprised to see the medium confined to a narrow cultural niche.*
>
> *(Ryan 2002 605)*

Though Ryan addresses the authors of hypertext, her words could easily apply to any genre. Theatre is already slipping out of the public sphere, often unable to capture the imagination, intellect, and attention of audiences now accustomed to the swiftness of digital interconnection and vivid images of modern media. At the same time, its audience remains larger and arguably more varied than that of performance art, where most critical performative explorations of technology are occurring. Can combined technique expand the audiences of both? Purely conceptual art is important, but over as soon as it is understood; performance art and 'traditional' theatre can strive for both concept and work itself, and bring motivated, critical use of technology to works accessible on several levels:

> *There is no contradiction in assuming both*
> *(a) that one must appreciate the whole structure of a work as the declaration of a poetics, and*
> *(b) that such a work can be considered as fully realised only when its poetic*

project can be appreciated as the concrete, material, and perceptibly enjoyable result of its underlying project… [T]he work fulfils its fullest aesthetic value only insofar as the formed product adds something to the formal model (so that the work manifests itself as the 'concrete formation' of a poetics). The work is something more than its own poetics (which can be articulated also in other ways), since the very process by which a poetic model acquires a physical form adds something to our understanding and our appreciation.
(Eco 1989 176–177)

What, then, does the digital offer to the collective experience of performance that can be incorporated into a dramatic text? Clearly one of the most intriguing and powerful capabilities of digital technology is its ability to support individual access and input, and we struggle to reconcile this idea with the structure of the dramatic text. Story branching, though well-supported by digital technology, is difficult to use effectively to show or suggest multiple perspectives. One apparent reason is that the audience remains unaware of the field of possibilities from which choices can be made, as those exist within the play's world, which cannot be collectively re-read, skimmed, or accessed out of performance sequence.

In our *Iliad*, the lives and world of the audience provide that field of possibilities: one of which each audience member has intimate prior knowledge. We construct a re-arrangement of their details for the purposes of the characters, with the belief that the story will open into new and resonant interpretations through each audience, through the small fibres connecting individuals into the story by digital technology. In Figure 4, once choices are made in performance between story sections (Figure 4A), the audience sees only the 'collapsed' structure (Figure 4B) and is not aware of or able to access other paths of the story. If external information is used from the world of the audience (Figure 4C), though the collapse of possibilities occurs when a particular choice is acted upon, their prior

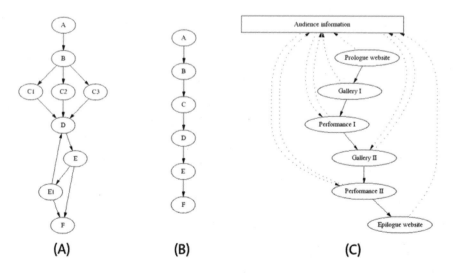

(A)   **(B)**   **(C)**

Figure 4. Branching story structure, and the audience as a source of information for non-branching narrative.

knowledge of and engagement with the field of possibilities allows them to see the choice within its context and, in our story, to come to recognise the process of cut-up that the Achaeans (and on another level, the authors) use to create and manipulate their experience.

Little about the details that are incorporated into the performance can be predicted, so their presence and interrelationships must be justified within the world of the play with only basic knowledge of content; the political machinery of the play's characters, which reuses anything it can find for its own purposes, addresses this consideration. At this stage in our struggle to incorporate digital technology intrinsically into a dramatic text, we create the necessarily absurd but absolutely self-consistent logic of the play's characters and their causes, which are both dependent on and independent of the specifics of the audience, and this brings us back to the current cusp of modern drama. Though Brechtian in its reference to reality, this re-use and re-arrangement of real-world specifics moves beyond a purely didactic technique because manipulation of information at the abstract level of computer software allows and even encourages the real world's technological processes to be amplified, distorted, and extended into the particular logic of the performance—the domain of absurdism.

Absurd story structures, such as Ionesco's, were already deeply rooted in the evolution of technology and often referenced the machinery of political power, power that has always been constructed upon its own self-consistent logic and grows further in distance from reality as its bureaucracy increases. A similar intrinsic and almost meglomaniacally self-consistent logic can be found in many complex technologies, and both are embodied in the media systems that implement our *Iliad*. The authors of modern drama were reacting to the Industrial Revolution, World War I, and World War II, as we react to today's mechanisms of power in technology and media. Alfred Jarry's *King Ubu* already amplified, distorted, and extended the world's processes and trappings of power, and following it there is clearly a path of reactions to the world's altering perspectives as the development of technology exploded. *Waiting for Godot* stripped away technology and politics, yet paradoxically showed how these factors have affected man's psyche. Like Jarry with *Ubu*, Beckett portrays the brutality of the human spirit, and by maintaining a celebratory tone, his characters are allowed to vaguely allude to profound philosophical questions through a parade of vague specifics, never truly answering a thing. Ionesco's characters in *The Chairs* attempt to tell the world of their total human experience, yet nobody is there to listen. Absurdity's construction of completely logical illogic continues to shine through all forms and genres of dramatic literature, including our own, as they continue to react to a changing society.

Digital technology allows these elements of modern drama to be drawn together in new ways: it can embody and execute 'logical' processes and simultaneously blur the boundaries between specifics constructed within the play's world and elements from the real world of the audience that are gathered and recontextualised dynamically. A reaction as much to the role of technology in modern society as to its potential impact on the dramatic text, our *Iliad* moves between the gathering of information (e.g. overtly on the website, covertly in the galleries) and the dynamic and absurd re-presentation of content customised from the 'demographics' of the audience. Lev Manovich finds a similar "oscillation between interactivity and illusion" to be a structural feature of modern society:

> While modernist avant-garde theatre and film directors deliberately highlighted the machinery and conventions involved in producing and keeping the illusion in their works—for instance, having actors directly address the audience or pulling away the camera to show the crew and the set—the systematic 'auto-deconstruction' performed by computer objects, applications, interfaces, and hardware does not seem to distract the user from giving into the reality effect. (Manovich 2002 208)

He goes on to define a new 'metarealism' that contains its own critique inside itself:

> Metarealism is based on oscillation between illusion and its destruction, between immersing a viewer in illusion and directly addressing her. In fact, the user is put in a much stronger position of mastery than ever before when she is 'deconstructing' commercials, newspaper reports of scandals, and other traditional noninteractive media. The user invests in the illusion precisely because she is given control over it... The oscillation analysed here is not an artifact of computer technology but a structural feature of modern society... This may explain the popularity of this particular temporal dynamics in interactive media, but it does not address another question: Does it work aesthetically? Can Brecht and Hollywood be married? Is it possible to create a new temporal aesthetics, even a language, based on cyclical shifts between perception and action?

The power structure that exists between author, actor, and audience-participant within this new type of dramatic work mimics the society and the new media that Manovich describes. Our *Iliad* tries to find a structure that uses this oscillation between individual interaction and performance to allow the collectively experienced event (for 'mass' consumption) to intrinsically incorporate specific information about the audience members (demographics) into the action.

Glimmers of this technique exist already in performances that use live images of the audience; consider the presence of a single live video camera selectively pointed at the audience, with its images shown on stage. If the live image's presence is justified by the story itself and is not solely a design or directorial brushstroke, the same dissolution of boundaries between the play's world and reality occurs, without disconnection from the characters and their logic, a feature of the metarealism described above. If the justification extends beyond a moment, and is intrinsic to the whole of the experience—if it, like the details of Ionesco and Beckett, arises naturally from the play's world and is utterly necessary in order to achieve its ends—then the presence of the audience, recontextualised by technology, might perhaps play a role in some new type of open-ended catharsis.

To achieve this in the example of a single live camera, one can imagine that the text's author must provide instructions for when to turn the camera on and off, where to point it, as well as creating a story that encompasses the presence of the live image, regardless of what it shows. In a manner different than a film's shot list, the author might suggest where to point the camera and when, even without being able to control what would be on camera at that moment. If these rules were codified, executed automatically by machinery programmed with when and where to capture and re-display images, the number of cameras and sensors expanded, and the resulting images were combined with other information gathered throughout the experience, placing fragments of the audience's real-world lives into the midst of the play's action at a speed mimicking the real world's networks of information, we arrive at our *Iliad*.

In it, we explore an arena that is built more on processes than specific media: modern processes of media recombination are incorporated in the script, which deals in functional relationships between information, as in the example described earlier of the creation of geographic boundaries referenced in performance based on the addresses of the attending audience. Rules and algorithms built into new dramatic literature can define the gathering (when and what to capture), manipulation (how to alter and juxtapose it) and delivery (re-presentation) of information in the audience's world to achieve dramatic goals in the plays, while leaving open the possibility for (re)interpretation by directors and designers of many input and output details. Outside of drama, the use of carefully chosen but seemingly arbitrary 'specifics' gathered from the audience is also not without precedent:

> *The contemporary novel has long tried to dissolve the plot (here understood as a sequence of univocal connections necessary to the final denouement), to construct pseudo-adventures based on the 'stupid' and inessential facts. Everything that happens to Leopold Bloom, Mrs. Dalloway, and Robbe-Grillet's characters is both*

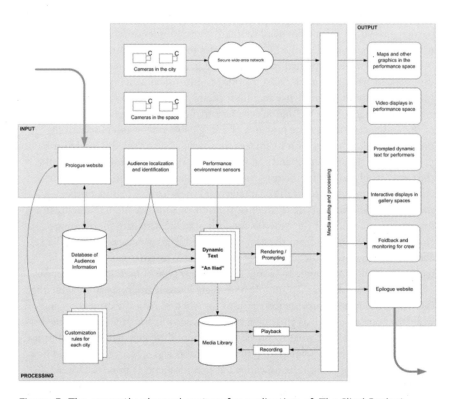

Figure 5. The currently planned system for realisation of *The Iliad Project*.

'stupid' and inessential. And yet, looked at from a different narrative standpoint, all their experiences appear quite essential to the expression of the action, to the psychological, symbolic, or allegorical development that implies a certain vision of the world. This vision, this implicit discourse that can be understood in a number of ways, and that results in a variety of different and complementary solutions, is what we call the 'openness' of a narrative work: the rejection of a plot signifies the recognition that the world is a web of possibilities and that the work of art must reproduce this physiognomy.

Whereas theatre and the novel have long been progressing in this direction (I am thinking of works such as those by Ionesco, Beckett, and Adamov, and The Connection, *by Gelber), cinema, another art based on plot, has been shying away from it.*

(Eco 1989 114-115)

The potential absence of plot in a new drama that clips and reuses information about its audience and their real-world surroundings should not be misinterpreted as a lack of story. If anything, the story itself takes on a more important role, being comprised of a clear series of events that are choreographed to lead to the finale. As with the works Manovich mentions, only glimpses of relevance may be perceived during the course of the story, and perhaps taken as inessential additions at first—the initial sight of one's image on a screen may only evoke a gasp, giggle, or grimace—but the final action can now draw upon the audience's exposure to continual re-contextualisation of information from their own lives over the course of the entire set of events, yielding a conclusion with open-ended and self-reflexive relevance.

Although Beckett and Ionesco point out the human errors of putting too much faith in technology, or any construction for that matter, the advancements that were occurring during their times shaped their work. Einstein's theory of relativity shattered the modern world and thrust the planet into a post-modern conundrum. Today we exist against a continual backdrop of technological advancement that seems to grow in autonomy and ability to re-present our own images back to us, as the Achaeans (and later, the Trojans) do to the audience of our *Iliad*. From health to democracy to thought, what is available to us comes via technology, and our own contributions to these are input via technology. Yet as the lists of choices grow, the sources of these choices become fewer and fewer, and with every touch of the mouse or the switch of a remote control, we receive updated, personalised stimuli. Every product we purchase is scanned and used for the purposes of marketing to those who are purchasing. In some ways, however, we are a little luckier than Beckett and Ionesco—the technology that shapes our world can be used on the stage, and we can construct our drama around actual processes instead of proxies, incorporating the audience within these systems' absurd extension of our world.

References

Birringer, J. (1999) Contemporary performance technology. *Theatre Journal* 51(4) 361–381.

Burke, J. (2002) Dynamic performance spaces for theatre production. *Theatre Design and Technology* 38(1) 26–34.

Eco, U. (1989) *The open work.* Harvard University Press, Cambridge, Massachusetts.

Manovich, L. (2002) *The language of new media.* The MIT Press, Cambridge, Massachusetts.

Reaney, M. (1999) Virtual reality and the theatre: immersion in virtual worlds. *Digital Creativity* 10(3) 183–188.

Ryan, M.-L. (2002) Beyond myth and metaphor: narrative in digital media. *Poetics Today* 23(4) 581–609.

Saltz, D. Z. (1999) Beckett's cyborgs: technology and the Beckettian text. In Schrum, S. (ed) *Theatre in cyberspace: Issues of teaching, acting and directing.* Peter Lang, New York, pp. 273–290.

9 Haunting the movie: embodied interface/ sensory cinema

Toni Dove

What is the impulse that compels us to sit in the third row in a movie theatre, pushing at the physical boundary between the body and the screen, between the material world and the immaterial world of representation? A close-up relationship to the screen fuses the presence and immediacy of theatre to the flickering light of cinema—we look for ways to 'really be there'. The realm of responsive media has opened doors for me into the possibility of an immersive sensory experience that can transform narrative structure. As if the words 'once upon a time' could become a physical passageway into the world of story transformed into architecture.

Current iterations of the action movie have an extreme physicality motivated in part by an economy of 'dumbing down' in the film industry. Global markets require minimal cultural specificity and reduced narrative complexity leading to a focus on pyrotechnics. The action movie's use of the eyes to create a roller coaster for the sensorium suggests to me alternate paths through narrative experience that could utilise this physical vocabulary to spatialise experience and explode some of the tidier conventions of the commercial film plot while enhancing subtext and complexity. In other words, visual cues can be expanded to include the entire body as a haptic interface creating a more nuanced somatic connection to narrative experience. This involves a dismantling of narrative and cinematic language to arrive at a different method of story construction. For me this involves a multiplicity of points of view (POV) positions triggered by an embodied interface experience that tugs against time by creating a complex horizontal narrative space. I have a continuing impulse towards the subversive abuse of technologies and established syntax—how can I use something in spite of, or against, the way it was designed? What if the story was a virtual building, what if we cut an intimate flirtatious interaction like a car chase. What if…? What if…?

The projects outlined that will act as examples here, *Artificial Changelings* and *Spectropia*, are the product of a decade of development in responsive narrative. They will serve to help me articulate some ideas about embodied interface, remote agency and telepresence and the merging of performer and audience within the context of interactive or responsive environments. How are new technologies rewriting our concepts of the body and its agency? If traditionally passive viewing becomes active how does this effect our definitions of character, story, performance and audience?

Artificial Changelings: the cut vs. the flow

Artificial Changelings is an interactive narrative installation that uses video motion sensing to track the location and movements of a viewer standing in front of a rear projection screen. A fragmented portrait, an exposition engine, a romance thriller about shopping, this interactive movie follows the life of Arathusa, a kleptomaniac in 19th century Paris during the rise of the department store, who is dreaming about Zilith, an encryption hacker in the future with a mission. Arathusa, the kleptomaniac, became a lens through which I could examine a subjective view of the effects of the evolutions of consumer culture (in this case the birth of the department store and with it the new strategies of seduction and display) on the individual through the disorders it engendered. I was viewing neurosis as a character armature and as an involuntary strategy of social resistance. Zilith, the encryption hacker and Arathusa's dream of empowerment, became as well a vehicle to examine the evolving technologies of the internet and their incorporation into consumer culture. The characters and their contexts were ultimately linked by the premise of shopping and its apparatus for the construction of desire in each century. I attempted to form each character in tune with their century and the constructions of identity associated with that period. Arathusa has a challenged yet coherent interior life and a single voice, while Zilith has no clear exterior or interior and often manifests as a polyvocal chorus. These constructions are enhanced and elaborated through the viewer's physical engagement with characters using body movement.

In *Artificial Changelings* a viewer's physical movement while standing in front of a screen moves a character's body, generates speech from a character, and alters or creates the soundtrack. In addition, floor pads act as triggers to dissolve between different video spaces organised according to levels of intimacy with the character. Standing close to the screen puts you inside a character's head (close-up), the middle pad puts you in direct relation to the character—as in conversation (direct address), the far pad is a trance or dream space—an altered state (trance/dream). A smaller pad behind the main three floor pads is a time tunnel that allows the viewer to move back and forth between the two centuries of the 'story' at will.

The two elements critical to the experience of the piece for me were:
1. the ability to create random access to multiple video clips and to segue between sections with dissolves, and
2. the transparency of the interface for the user.

In other words I wanted no apparatus, visible hardware or devices to intervene between the user and their experience of agency (their ability to make things happen) in the installation and I wanted the experience to feel like an uninterrupted flow. These elements are what contribute to the hallucinatory nature of the piece. They create a physical fluidity, an experience different from that of film, which uses the cut as the architecture of time and space. There may still be cuts used as an editing device, but the centre of the experience is characterised by

Figure 1. *Artificial Changelings*. Installation still with viewer interacting in the piece. Art Center for the Capital Region, Troy, NY. ©Toni Dove 1998.

unbroken flow—a sense of continuous motion connected to the body's sensorium, the somatic experience of perception. In this context a badly placed cut becomes a disturbing physical experience. The awareness on the part of the viewer of being simultaneously in their body and in the screen is key to the immersive experience. The space between the body and the screen becomes charged.

The architecture of attention

'Physicalising' virtual space with the body in motion suggests the need to re-examine not only how we define the body's boundaries, but also the structures of attention that have been defined through passive viewing. The definition of attention as a single stream, a focused, aimed experience that is cerebral, localised and specific may be turned on its ear by the surfing repetitions of responsive interface. The repetition of physical action, the sense of attention being almost the background or field that allows another form of experience to surface is similar to the way one's mind wanders while driving a car—the familiar almost invisible repetitive actions release the mind. These repetitions of action form a trance-like experience and so it becomes useful to look at trance, meditation, inattention, boredom and hypnagogic states in relationship to how experience is constructed in responsive media spaces. It is notable that the word 'surfing' has become so common in relation to Web experience—this idea, of coasting across data and information in a fluid and malleable way driven by the user, has an experiential relationship to swimming. This is a different form of attention—a kind of sustained tension that creates a space for reception. It is a series of vertical eruptions in the horizontal field of time. Information leaches up through layers or strata to create meaning and is a departure from the temporal linear sequence of conventional plot. I am interested in working with a form that I would call 'coherent channel surfing'—organising and structuring the nomadic random access of data navigation and using it to design specific narrative databases.

The feedback loop

Gesture is often thought of as a discrete entity—a physical movement in time with a specific temporal shape that registers almost as an object—a repeatable sign. If we think of gesture as a generating or controlling device in interface, we miss what have become for me the most interesting possibilities inherent in physical motion. During the development of *Artificial Changelings* I noticed that viewers went through a process in their relationship with the piece. They initially tried to control the character with actions and gradually slowed down and gave in to the system and to the character, often mirroring the character's actions the way human beings mirror each other's body language in conversation. The motion sensing system in *Artificial Changelings* parses motion below the level of perception. A gesture is not an object or sign to produce again and again as a discrete whole but a series of changes in time too complex to repeat. What appears to be the same gesture will not repeat the same result—the motion sensing system maintains a connection to the viewer that is not quite predictable. A basic set of simple operations in the interface are quickly learned (speed or amount of motion, motion to the left or right as well as movement back and forth on the floor pads). A wide range of possible content is supplied in response with a high degree of unpredictability of behaviour in the results—again not unlike the experience of conversation. The viewer's body motion is tracked in terms of the spatial location of motion and the amount of motion. While viewers may choose to mirror the actions of a character's gestures or not—it is the flow, speed and location of motion rather than any particular semiotics of gesture that connects the body to the media. This constantly changing feedback loop between the body and the media affects the way agency and gesture play out in performance—sometimes motion generates action; sometimes motion follows action—until the difference disappears in a continuous flow: the feedback loop. The body is stuck to the movie, a part of it, lost in space and time. This effects the way a viewer moves, and perhaps how we might think about what a body is—its boundaries and edges go soft.

   The viewer's experience of their body in the screen, (the telepresent agency that leaves traces of their actions in real time in the media that comes back to them), mixed with the physical repetitions of motion in time, is what creates a sense of immersion in the responsive media space. This combination of action and physical sensation induces a trance-like state physically connected to the media that contributes to spatialising the narrative experience and disrupting a linear or sequence based notion of plot. My current project *Spectropia*—a work in progress—adds voice to the authoring system and interface used in *Artificial Changelings* and brings it more into the realm of performance. It also extends the duration of the experience and is built around a central narrative arc.

Close-up

S
P
E
E
C
H

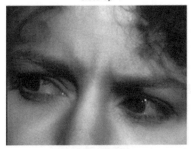

I'm sick of waiting!
I'll stop!
I won't be bored!
Too much tension!
I'm so bored!
I'm restless!
Stop time!
Slow down time!
Speed up time!
Seize life now!
Don't wait!
Beautiful Dreams!
No more tedium!
More waiting!
Waiting for what?

I'm so tired of waiting
Always waiting for something
Always waiting for someone
The day goes by so slowly
The days rush by
Time stretches endlessly
Time in front of me
Time passing by me
Morning toilet, lacing, dressing
Visits and calling cards
Shopping and tea
I 'm restless and nervous
Tension mounts in me
I'm not quite myself
I dream strange dreams
I see spectral appearances
Time seems to slow down
My perceptions are altered
Boredom is so dreadul
What am I waiting for?

Take it!
Put it back!
I want it!
It will be mine!
It's mine!
I don't need it!
I won't take it!
The suspense!
The thrill!
I't's gambling!
No more!
No more waiting!
More freedom!
Gambling!
Don't think!
I'm possessed
What am I doing?

I stroll down the aisles
It's as if everything belongs to me
I might take it all.
It is like gambling
Games of chance - suspense
It frees me from waiting
Reasoning powers are suspended,
A burning mounts to my brain,
Nothing remains but suspense
The thrill of it
Like a drug for time
All the senses are engaged
I would do anything
I do not think of what I'm doing
I can't control my hands
I'm possessed of an impulse
I seize things and walk off.
What goes on in my frenzied brain?
It is the casino of commerce
A theater of dream and novelty

T
r
a
n
c
e
/
D
r
e
a
m

Figure 2. *Artificial Changelings*. Navigating the three zones of a scene with Arathusa, the character in the 19th century. A viewer's motion generates speech and gesture from a character. ©Toni Dove 1998.

C
l
o
s
e
-
u
p

D
i
r
e
c
t

A
d
d
r
e
s
s

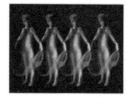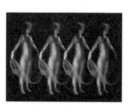

T
r
a
n
c
e
/
D
r
e
a
m

Figure 3. *Artificial Changelings.* Navigating the three zones of a scene with Zilith, the character in the future. A viewer's motion generates movement from a character. ©Toni Dove 1998.

Spectropia: cinematic Bunraku

Spectropia is an evening-length interactive media performance. It will be performed by two players on multiple screens for and by an audience. A version of *Spectropia* is also planned for home interaction using a combination of DVD ROM and internet delivery. I am also considering producing a linear movie version. *Spectropia*, shot on digital video, is a time travel drama set in the future and in 1931 after the stock market crash. It uses the metaphor of supernatural possession to explore new constructions of subjectivity and the disorders brought on by consumer culture and emerging technologies. Who is in there? How did they get there? Who put them there? These questions are for me a research adventure navigating the tug of war between the social and the individual; between the markets determining culture and the voluntary as well as involuntary strategies of resistance we employ against them without always being fully conscious of our actions. How we are determined and how we fight back. The narrative forms—the propaganda recipes of our entertainment industries—become an interesting critical tool on this voyage, both to use and to subvert and disassemble.

Synopsis

The story opens in a time reminiscent of the late 18th century, but strangely inverted, as if seen through a looking glass of dimly remembered and mutated images. We are in the future in England, 2099, a world of artificial surfaces where knowledge spans only a person's experience and recorded history is forbidden. This culture of consumption floats on islands of garbage; saving anything is punishable by law.

Spectropia, a self-taught archaeologist in her early twenties, lives in the salvage sector of an urban centre known as the Informal Sector. The Sector's function is to compact trash in order to lift the island above sea level as the water slowly rises. She seems to live, partly by choice, in an environment devoid of human presence. *Spectropia*'s companion, a duck android (part animation and part wireless robot) runs a black market business in retro objects. The Duck is a babysitter bot, *in loco parentis*, programmed by her father who disappeared while searching back in time for a missing inheritance. Driven by a feeling of abandonment, *Spectropia* searches for clues with a scanning machine of her own invention. The machine reads garbage and translates it into lifelike simulations that she can enter and that respond to her voice and movement. This illegal research into history takes her to the 1930s in New York City after the stock market crash. Obsessive scanning short circuits her machine and traps William, a handsome ghost, in her workshop. He is a black and white phantom, caught out of the simulation and superimposed over the baroque trappings of the future.

Spectropia's machine transports her to William's time and place, New York City 1931, where she finds herself in the body of Verna DeMott, a sophisticated older woman and an amateur sleuth. She inhabits Verna incognito, and helps William in his search to recover money stolen in an investment fraud that has destroyed his reputation. A strange romantic triangle involving multiple personalities in singles bodies, time travel, and a mystery involving possession take the couple on a journey—clues appear written in invisible ink, hidden in locked boxes, broken lockets, at the end of dark tunnels and winding stairs, in towers.

Spectropia uncovers the mystery of time travel on her machine: clothes are the passageways between the centuries. A garment slips off her shoulder as erotic attraction builds and 1931 trembles, dissolves and fades to the future. She is yanked forward in time only to scan herself back again to finish the puzzle. The tension between repression and release drives the narrative forward to its climax, in which economic disaster is only partially averted, romance is both thwarted and rewarded, and *Spectropia* returns to the future with a new understanding of her own past.

The interface and narrative

Unlike traditional movies, *Spectropia* is 'performed' interactively (by trained performers and ordinary viewers) using a mix of motion sensors, speech recognition and synthesis, and vocal triggers. Viewers assisted by trained performer tutors can spontaneously unfold dialogue between characters based on physical cooperation, speak to characters and have them respond, navigate through cinematic spaces, move a character's body, and alter and create the sound design.

Spectropia's performance interface mirrors its theme of otherworldly possession, as 'viewer/performers' inhabit characters in an echo of *Spectropia*'s experience in the body of Verna in 1931. The cinematic action is projected on a central wide movie screen, flanked by smaller screens that allow cinematic characters to break from the screen and move onto the stage to interact with performers. Unlike a film, however, these projected screen characters are 'inhabited' by two viewer/performers who use movement, vocal sound and speech to manipulate them like virtual avatars. These viewers can make their on-screen characters move through suspense-filled action, alter and inhabit dialogues, engage in intimate interactions (karaoke flirting), or speak with virtual characters in the form of simulations from *Spectropia*'s machine. Suspense becomes a space to be manipulated—stretched and compressed in time. The emotional impact of dialogues is changed by alterations in pacing and speed. Occasionally the screen splits and we choose to follow one character or attempt to track both. Visual and audio cues tip off the audience as to how each viewer is shaping the onscreen action. *Spectropia* might be considered an advanced technology variation on Japanese Bunraku puppet theatre, in which shadowy black-clad puppet masters perform on stage, articulating nearly life-sized puppets. The shadows of agency are embodied and visible on stage.

Spectropia's trained tutor/performers are more than just mute puppeteers though; they may appear onscreen themselves 'haunting' the movie, cast shadows into the film world, or comment directly to the rest of the audience and act as guides so that members of the audience can enter the movie. Their movements also contribute to creating a spontaneously composed interactive soundtrack.

Subverting the standardised story

Stories are the propaganda organ of a culture. They tell us how to live and how to think about who we are. They operate to transmit the dominant ideology of the mainstream of culture, but are available to be bent and altered by an audience willing to examine the nature of the language. An excellent example of this are the re-edited versions of *Star Trek* created and circulated by fans that address issues of gender and sexuality not intended in the original broadcast television show. This is both an audience intervention and a form of interactivity generated in part by inexpensive VCRs and editing systems. When new technologies appear that require new syntax for delivery there is an opportunity to transform or make interventions into a language that has tremendous impact on culture.

Taking the recipes for the suspension of disbelief present in genre fictions and fusing them with the suspension of disbelief recipes present in interface seemed

Figure 4. Performance still from a demo of *Spectropia* at the Institute for Studies in the Arts, Arizona State University where the prototype for the project was created. *Spectropia* ©Toni Dove. Work in Progress 2003.

like an interesting place to start re-thinking narrative and cinematic language. In this case, the ghost story of supernatural possession fused with the telepresent interface—the viewer/performer haunts the movie in a way that echoes the characters in the story haunting each other across time and space. The behaviours designed for the interface and the viewer's interventions into the media space blow holes in the linear sequence of plot, spatialising the experience and de-stabilising it through random access. It is as if an architecture of perception were being constructed around a linear story line composed of a multiplicity of POVs designed to make the viewer feel themselves in their own body. The design of the behaviours derails the tropes of genre and allows them to be triggered in an almost free-floating context. A flashback button that loops the viewer's experience in time, reviewing fragments of previously viewed or missed moments, exaggerates this exploded experience of genre. These devices act simultaneously as anchors into a familiar narrative space and as the disruption of that space. The audience, intimate beyond its own conscious knowledge of every rule and detail of genre, can take a collection of familiar but fragmented tropes and fill in missing information. Narrative becomes an ambience, a place as much as an engine moving through time. The body's connection to the media enhances the hypnotic nature of the experience and creates glue for the continuous and immersive environment. The movie falls apart and re-assembles itself, wrapping around the viewer.

In designing behaviours for *Spectropia* I was thinking about the way perception is experienced as actual physical sensation in the body. I wanted to create a virtual architecture that would enhance the psychological subtext of the story through physical experience. It involved deconstructing cinematic language and examining the physicality of perception. What I ended up with was an interface that gave two performers the ability to inhabit and accompany two characters through a virtual space. The viewer's somatic and vocal relationship to the virtual space and to the characters and the characters relationship to each other and to their own interior space construct the architecture of immersion. The viewers communicate to each other through the characters on the screen—they haunt the movie. In the performance version the viewer/performers play the movie instrument for an audience. Pushing *Spectropia* into the arena of performance brought me back into the realm of the passive viewer. If there is an audience that doesn't participate in the action, how does the design of the interface and interactivity impact on what the audience sees and hears? Can the movie instrument play the audience?

Spectropia is for me a research project that examines the granulation of narrative cinematic experience. I wanted to break the movie down until it was like sand—to the point where you could push on it and change its shape. The behaviours of the program and the evolving feedback loop with the viewer in the piece generate changes in sound and picture that reflect the body's relationship to the virtual space, the characters' relationships to each other, and each character's interior voice. While this feedback loop of motion between body and media is a sensual experience for the active viewer, the resulting visual and sonic experience can bring that sensuality back to the passive viewer. This is another feedback loop—from the loop between body and media to the resulting media creating a sensory response in the passive viewer. This is an experiment for me—to see if the behaviours can work to enhance narrative in a way that will create both a convincing experience for the audience of the performance (aided by what I call 'performance transients'—moments of theatrical action by the performers that articulate the function of the program behaviours and the interactivity) and for the viewer of a linear version of the material—the movie. Backing into cinema through a side door to be surprised into re-seeing it in a new way.

I hesitate to claim this as a 'new' form of passive visual cinema, but I am trying to discover for myself a way of breaking loose from some well-worn cinematic conventions. I can see this attempt to form a language as a line drawn from Eisenstein through Hitchcock to the present. Eisenstein used the cut to elaborate subtext in a way that D.W. Griffith did not. Griffith is the grandfather of our current commercial narrative cinema language—it is a tidy cinema of narrative resolution. Eisenstein's desire to elaborate subtext through the rupture of the cut is, I think, present in Hitchcock's visual language that privileges psychological subtext over story line. An example is Hitchcock's famous concept of the 'McGuffin'. The thing that everyone in the story seems to care about most, but which is, in fact, only a device to allow for a love story or another set of relationships to emerge. A device to elaborate subtext. His extreme camera work using distorted angles and exaggeration (think of the staircase in *Vertigo*, 1958) or the constructed visual metaphors (the kiss scene with the sequence of opening doors in *Spellbound*, 1945) serve this concept—it represents emotional and psychological states. I am looking beyond the car chase into a physical language that enhances narrative subtext using physical sensation to articulate emotional and psychological space.

Figure 5a. Spectropia conjures Sally Rand on her machine. *Spectropia* ©Toni Dove. Work in Progress 2003.

Figure 5b. Sally speaks: viewers talk to Sally with a microphone and she responds. The program uses speech recognition and synthesis. Spectropia, ©Toni Dove. Work in Progress 2003.

Figure 5c. Sally's dressing room: viewers name objects and they speak and sing. *Spectropia* ©Toni Dove. Work in Progress 2003.

Figure 5d. Sally's bubble dance: viewers movement re-animates Sally's bubble dance while their voice alters and creates the soundtrack. *Spectropia* ©Toni Dove. Work in Progress 2003.

There is an erosion in responsive media of the boundary between audience and performer—another potential feedback loop that challenges the 19th century convention of the passive audience. It is interesting to think about the ways in which responsive media can impact on public social space rather than retreating entirely into the private space of one person, one computer. One way of approaching social spaces in the private sphere is through multi-user-networked communication, but approaching the problems of public space with responsive media is a different challenge. It requires addressing the convention of the passive audience and also grappling with the issue of large groups experiencing something responsive. This has led me to examine the idea of the performed instrument as a model. Performers become the virtuosos of the movie instrument for an audience. *Spectropia* is for me a way of examining this boundary between audience and performer and of taking it apart and pushing on it from several different angles. The players in the performance version of *Spectropia* play the movie instrument for an audience, but they may also walk into the audience and engage audience members in the movie, allowing them to talk with interactive characters or even enter the performance zone. I would like to workshop versions of *Spectropia* to look at audience behaviour, experimenting with the agency of the performers and with the breakdown of the barrier between the two in different contexts. The performance version can become a tutorial for the desktop version. In the desktop version, as in interactive installations or games, the viewer becomes the performer, inhabiting or participating in the virtual environment from the gallery or the privacy of their own home, perhaps linked by network to a friend in another location.

I like to think of my projects as philosophical toys with multiple personalities—iterations in public space, private space, linear form, and responsive environment. This work for me is a research project that began with building the instrument and is now taking me towards a re-examination of forms of reception and distribution from the living room and the theatre to marketing strategies for distribution. It is a journey through the architectures, virtual and material, aesthetic and economic, that determine the platforms of art and entertainment. In the end I would like to be able to create an intelligent and accessible entertainment that did not underestimate its audience and could support itself financially. It seems like a modest desire, but turns out to be quite a challenge. I think it's an important model to establish—a niche market that could function as a research resource for the larger culture, protecting and encouraging spaces for experimental practice. And this brings me back again to the subversive abuse of technology—how to subvert standardised production and reception recipes and bend them to new uses? If stories are the propaganda organ of the culture then I want to participate in some small way in the development of that syntax. We need a rich language for cultural propaganda so that we can bend it with a chorus of voices.

Artificial Changelings debuted at the Rotterdam Film Festival, 1997–98. The piece was supported by grants from the National Endowment for the Arts, The New York State Council on the Arts, The New York Foundation for the Arts, Art Matters, Inc., Harvestworks, Inc. and the Eugene McDermott Award in the Arts from M.I.T. It was part of the exhibition *Body Mécanique*, at the Wexner Center for the Arts, Ohio, from September 18, 1998 through January 3, 1999. A Catalogue was published with the exhibition. *Artificial Changelings* was shown at the Institute for Studies in the Arts at Arizona State University in March 2000 at the International Performance Studies Conference and was part of the exhibition *Wired* at the Art Center for the Capital Region, Troy, 2000, part of the *Book-Ends* Conference. *Artificial Changelings* will soon be released as an interactive DVD ROM for mouse and microphone.

    Spectropia is a work in progress. *Sally or the Bubble Burst* is a scene from *Spectropia* adapted for single users on the desktop. It was released in 2003 and is available on DVD ROM from Cycling '74 (cycling74.com). A research fellowship from the Institute for Studies in the Arts at Arizona State University (ISA) provided creative, programming and engineering support to develop the prototype. *Spectropia* has received grants from the Greenwall Foundation, the Rockefeller Foundation, the New York State Council on the Arts, the National Endowment for the Arts, the New York Foundation for the Arts, the Langlois Foundation, and the LEF Foundation.

References

Augaitis, D., Macleod, D. and Moser, M. A. (ed) (1996) *Immersed in technology: art, culture and virtual environments.* MIT Press, Cambridge, Mass.

Bergson, H. (1999) *Matter and memory.* Sixth printing. Zone Books, New York.

Carter, C. (1999) *The art of the X-Files* (based on the X-Files created by Chris Carter). Introduction by William Gibson. Harper Collins, New York.

Eisenstein, S. (1997) *Film form.* Harcourt Brace, New York.

Hill, L. and Paris, H. (ed.) (2001) *Guerrilla performance and multimedia.* Continuum, London.

Pearson, K. (2002) *Philosophy and the adventure of the virtual: Bergson and the time of life.* Routledge, London.

Reiser, M. and Zapp, A. (ed.) (2002) *New media narratives.* BFI Press, London.

Truffaut, F. (1983) *Hitchcock/Truffaut.* Revised edition. Simon and Schuster, New York.

Reisz, K. and Millar, G. (1977) *The technique of film editing.* 22nd edn. Hastings House, New York.

Films

Hitchcock, A. (1945) *Spellbound.* Selznick International, United Artists.

Hitchcock, A. (1958) *Vertigo.* Alfred Hitchcock, Paramount.

10 Play it again, Sam: film performance, virtual environments and game engines

Michael Nitsche and Maureen Thomas

Theatre and the virtual space

This chapter focuses on analysing the results of a studio workshop experiment in interactive improvised performance in a computer-generated and implemented environment, carried out in 2002[1]. While the value of real-time three-dimensional virtual environments (RT3DVE) for either theatrical or cinematic use has been evaluated in a number of contexts, live performance and its theatrical qualities in combination with the visual language of cinema has rarely been investigated. This vast territory remains largely unexplored, although it combines two of interactive media's most important features—live and immediate interactive access (as in theatre performances) and audio-visual mediation (as in cinema and television).

The graphical representation at the heart of a RT3DVE combines the theatrical performance with the cinematic interpretation of the evolving scene. Every RT3DVE has to be presented to an audience through some form of audio-visual mediation that demands designed camera work—only the camera makes it legible. As there can be no immersive visual RT3DVE experience without such consciously designed camera work, there can be no theatrical performance inside a virtual space without a cinematic dimension.

Virtual environments (VEs) for creating dramatic experiences have been discussed since the mid-eighties (e.g. Laurel 1986) and considerable attention has been paid to their cultural (e.g. Tomas in Benedikt 1991, Ludlow 1996), narrative (e.g. Murray 1997) and social potentials (e.g. Turkle 1995, Stone 1995). Interactive access is key to dramatic engagement in the computer environment, and Espen Aarseth has theorised this as a crucial part of a 'textual machine', consisting of computer, human user, and exchanged sign. He elegantly describes the constant 'performative' production of interactive text, with the audience influencing the unfolding action (Aarseth 1997 21). Aarseth concludes that an interactive dramatic event only comes to life through some form of physical participation of the human user. Interacting in a VE becomes a performance by the interactor, who plays with/against/to the environment and its dramatic content.

VEs for dramatic theatrical performances have been experimented with in the research environment, ranging from basic improvisation (e.g. Perlin and

Goldberg 1996, Hayes-Roth et al. 1994) through performances in Internet chat-rooms (e.g. Schrum 1996), to pieces in 2D visual chat-areas (e.g. Saarinen 2002), and acting rehearsals in 3D space (Slater et al. 2000), or staged re-interpretations of Shakespeare (Tosa and Nakatsu's 1998–99 *Romeo and Juliet in Hades*). Commercial applications have been installed in amusement parks (e.g. Pausch et al. 1996, Mine 2003)—but dramatic VEs are most widely known in computer games.

RT3DVEs offer a dramatic immersive stage space. Janet Murray theorises this aspect of performing (Murray 1997), and Manovich goes so far as to suggest that in RT3DVE "*space becomes a media type*" (Manovich 2001 251) (Manovich's italics): RT3DVE can transport, generate, and modify content elements while immersing users in a seemingly coherent space. As Poole expresses it: "In the cinema, the world is projected *at* you; in a videogame, you are projected *into* the world" (Poole 2000 98) (Poole's italics).

The RT3DVE *Casablanca Performance Experiment* aimed to generate a satisfying theatrical performance by exploiting the potential of a games environment informed by cinematic thinking and dramatic expertise. After an analysis of the field as theorised by the above scholars, participants experimented practically, combining creative skills in spatial design with narrativity and drama. A RT3DVE embodies spatial, graphic, acoustic and cinematic codes that offer a wide range of creative options, transcending the mediative possibilities of written text. Virtual set design, lighting, and spatial sound design are vital ingredients in a RT3DVE. As there are no physical restrictions to VE space, the design-vocabulary is almost unlimited. A wide expressive field for performers and designers is available.

Cinema and the computer

A RT3DVE lacks a 'natural' point of view because it is not subject to the usual limits of real space which govern physical camera and performer movements, so designers have to create a visual representation—a camera's point of view—of whatever is happening, as it happens. This mediation is presented as moving images, and therefore has to deal with many of the same issues as cinematic representation. Remediation—"the representation of one medium in another" (Bolter and Grusin 1996 339)—of cinema via RT3DVEs is achieved mainly through reference. According to Bolter, we read the visual representations of RT3DVE much as we read film. Manovich, who also regards cinema as the main influence on the development of digital media objects such as RT3DVEs, rightly stresses the "most important case of cinema's influence on cultural interfaces—the mobile camera" (Manovich 2001 79).

Modern computer games quote film camera work and visual strategies such as slow motion (*Max Payne* 2001), zooming in and out (*Metal Gear Solid* 1997), or cutting to a pre-defined view of (the virtual) space (*Resident Evil: Zero* 2002) (King and Krzywinska 2002 offer detailed discussion). Titles such as these, situated along the blurring borderlines between film and game, nurture a culture of cross-referencing between computer games and cinema. RT3DVEs are used during the pre-visualisation of film productions like Steven Spielberg's *Artificial Intelligence: AI*

(2001) (Lehane 2001) they create movie spin-offs, like *Tomb Raider* (2001) and *Final Fantasy* (2001) and they influence the aesthetic as well as the content of films like *Tron* (1982) or *The Matrix* trilogy (1999, 2003).

Real-time camera control in a 'cinematic' narrative VE presents a number of research challenges, which are addressed both in commercial R & D (e.g. *The Getaway* 2003) and academic environments such as MIT Media Lab (Drucker 1994, Tomlinson 1999) and at Microsoft Research and University of Washington (He et al. 1998). This work aims to perfect a computer-controlled automated 'camera operator' which follows basic cinematic shooting and editing strategies expressed as rule-based systems. The *Casablanca Experiment*, however, took an approach closer to that of the 'inhabited television' projects *Out of this World* (Benford et al. 1998) and *Avatar Farm* (Benford et al. 2000) based at the University of Nottingham, UK. These projects staged small games in massive-multi-player environments that were visualised by virtual camera operators active inside the computer-generated world. Their individual camera viewpoints were assembled by a live vision mixer for television broadcasting.

'Inhabited television' uses RT3DVEs as arenas for dramatic performance, in combination with elaborate human-controlled camera work, in order to deliver engaging TV which exploits traditional expertise, using a developing medium, the computer (for visualisation strategies see Craven et al. 2001). Benford and Greenhalgh explore virtual communities, structuring their drama around them, and engaging participants in games or partly predefined scenes. The *Casablanca Experiment* concentrated on an environment structured very purposefully to create engaging drama. We were interested in extending the expressive range of RT3DVE for dramatic performances in virtual places by combining the principles of architectural design with principles of narrative and dramatic design. The mediation—the virtual camera work—was crucial to this investigation.

The *Casablanca Interactive Performance Experiment*

The *Casablanca Interactive Performance Experiment*'s overall goals were to identify and formulate a method of combining live performance with cinematic visualisation in a RT3DVE, starting from a classic dramatic film scene. It was designed as an arena for the analysis of the expressiveness of RT3DVEs, allowing a direct comparison between traditional performance in cinematic media, and interactive media. The experiment chiefly focused on developing the audiovisual language of a RT3DVE itself. The task was not to copy a given scene in a RT3DVE, but to re-interpret it using the expressive means available in virtual environments.

The four-day experiment was conducted with two teams of four Cambridge University MPhil students, all familiar with the cinematic use of camera, editing, design, and dramatic staging. To enable meaningful mediation in the RT3DVE, the cast of the film *Casablanca* (1942) was modified to include an invisible additional player—the virtual camera operator who defines the camera strategy which mediates the dramatic *mise en scene* and the expressive movements of the visible character-actors. Each character is steered by a 'player', and the camera operator becomes another active 'player', whose performance is as essential to the drama as that of the character-actors. The virtual cinematographer determines the final picture on the screen.

The experiment was inspired by Michael Curtiz' 1942 black and white movie, *Casablanca*, itself based on the Burnett and Alison play *Everybody comes to Rick's* (1941, unpublished). *Casablanca* presented the necessary challenge for a dramatic performance with its focus on the expressive characters' performances in the context of a romantic love-story, instead of the frantic action-driven spectacle usually at work in a commercial RT3DVE. The scene chosen for the experiment is a dramatic turning point in the overall story. It is set at night in Rick's apartment above 'Rick's Café Americain' and sees Ilsa (played by Ingrid Bergman) visiting Rick (played by Humphrey Bogart), her former lover. Ilsa has evidently betrayed Rick a year ago, leaving him and vanishing without trace. In this scene, Ilsa desperately tries to get transit visas out of wartime Casablanca from Rick, who is a smooth black-market operator, in order to save her husband—Victor Lazlo's—life. She begs, argues, and even threatens Rick. But Ilsa fails, finally admitting that she is still in love with Rick. The scene is a melodramatic high point and leads to the crucial change in Rick's behaviour that drives the action to the end of the film. In the studio workshop experiment, the participants performed this scene, steering avatar versions of Rick and Ilsa in a RT3DVE, with a third player operating the virtual camera.

Settings and tasks

The experiment aimed both to test whether VEs are capable of creating dramatic experience on a par with that enjoyed by cinema goers watching classic film, and also whether a 'viewable' performance, whose qualities match those of traditional film productions, could emerge in RT3DVE. The workshop participants watched *Casablanca* before the experiment, minus the selected scene between Rick and Ilsa, in order to avoid any direct copying of the camera work, staging, and set design from the movie. Video recordings were made of the 'live' performances in the RT3DVE, and these were compared with the original film scene at the end of the workshop. The two groups of graduate students developed their own visions of the entirety of the scene, from set-design to *mise en scene*. They performed their different versions in RT3DVEs that they designed and created for the purpose.

The performance itself was delivered in a multiplayer game-engine on the local area network (LAN) in the Digital Studios building. The participants were asked to

* Develop their dramatic interpretation of this scene:
  o design the stage/set (shape, size, colour, context);
  o stage the event (*mise en scene*, actors' movements);
  o define the cinematography (editing, camera movements).
* Implement the dramatic interpretation in the VE:
  o model the virtual 'set';
  o import the modelled 'set' into the 3D engine ;
  o texture the model;
  o light the virtual 'set'.

* Rehearse the scene in VE, experimenting with the *mise en scene* of the two characters and the camera in real time.
* Perform the dramatic scene in real-time for themselves and an audience:
  o one participant controlling the Humphrey Bogart character;
  o one participant controlling the Ingrid Bergman character;
  o one participant controlling the camera view.

In this way, the roles of the director (Michael Curtiz) and the cinematographer (Arthur Edeson) in the pre-production phase of moviemaking were taken by the whole team as they decided the most effective *mise en scene*. They also shared the role of the editor of a traditional film in post-production, at this pre-production stage, by defining cut-in and cut-out points. The performance itself was then generated altogether in real-time, so the two players steering the avatars Ilsa and Rick collaborated with the player operating the virtual camera, to produce the most effective dramatic version of the scene for the screen. The planned performance was realised in the actual show, which worked very much like a live theatre performance including last minute changes, instant adjustments, and the influence of various factors of the delivery platform as will be described below. The soundtrack was taken from the original movie, allowing a comparison between performance in film and RT3DVE. Technically it is possible to provide a live audio link between the avatar characters and the people steering them, which would enable real-time voice audio transmission, but quality tends to be (2003) too low for subtle yet expressive dramatic purposes.

Effectively, the players were enabled to design a virtual set to support a virtual performance, and then improvise the performance itself on that virtual stage, with the help of a dramatic script written by expert dramatists and a voice performance given by great actors.

Technical specifications

Although there were more powerful systems available, Epic's *Unreal Tournament* game-engine was chosen for the experiment for four main reasons:
* good documentation;
* proven stability;
* future extension;
* sufficient multi-player capabilities.

The *Unreal Tournament* (Epic MegaGames Inc. 1999) engine has been in use since 1998—when the original *Unreal* game (Digital Extremes and Epic MegaGames Inc. 1998) appeared—and the capabilities of its editor are well documented by a wide base of enthusiasts. Constantly updated through new releases, it is one of the dominating commercial 3D engines. It ties into various third party programs (e.g. modelling programs); additional tools (e.g. code debugger, or texture reader) are often freely available (for details: http://www.epicgames.com). The editor and the engine both proved stable and bug-free.

All *Unreal* systems build on each other, so knowledge of any *Unreal* editor remains viable for future updates of the system. The *Unreal Tournament 2003* (Digital Extremes and Epic MegaGames Inc. 2002) editor offers new features but matches in basic functionality and interface design. Finally, this system offers the multiplayer-capability needed to realise a theatrical performance, which includes more than one character in a VE. The chief rival game-engine, ID Software's *Quake* engine, might have done just as well, but we had more in-house knowledge of the *Unreal* editor.

The *Casablanca* experiment was delivered on a local area network (LAN), but with minimal technical changes the performance could have been done over the Internet, with interactors participating from anywhere in the world.

Another useful feature of the *Unreal* system is its cross-platform performance and parts of the workshop, including the scene-rehearsals, were performed cross-platform on Mac and Windows PC. The engine is tailored to run on a wide range of computers, and optimised towards performance under different circumstances. We used low-spec PCs (750 MHz PIII) and Mac G4s for assembling the scene, while the final performance was delivered on faster consumer PCs, with GeForce 4 graphic cards. No specialised hardware was needed to run the workshop.

Using a commercial platform such as *Unreal Tournament* has the advantage of proven stability and giving immediate results without having to code a new system. The counter disadvantage is the inaccessibility of the source code. Although the *Unreal* system offers the potential to manipulate settings through its proprietary coding language, *Uscript*, the main code of the render engine remains hidden, and cannot be changed. This short intensive workshop experiment used the available system, with no additional programming. Other research projects that have utilised this engine but added code are the *Virtual Reality Notre Dame* project (http://www.vrndproject.com/) and some research in interactive storytelling (Cavazza, Charles and Mead 2001), as well as investigation into mnemonic spaces (Fuchs and Eckermann 2001).

Results

Importance of spatial design

For the *Interactive Casablanca Performance Experiment* the spatial definition of the stage was crucial. The environment is not a computer game level to be 'completed', but the animate expression of the drama's core elements—a virtual stage. As educated architects most of the participants were familiar with spatial design principles and highly aware of the expressive possibilities of space. The two groups of participants each designed, modelled, and imported their own virtual sets. There were no limitations for the virtual stage, which allowed both groups to move as far away from the production design of the original film as they wished. Both groups in fact chose to change the spatial structure from the original 'Hollywood Morocco' (Art Director: Carl Jules Weyl) to modern high-rise skyscraper-like structures redolent of the virtual cityscapes of Futurism and Manga animation.

There are no such elements in the original film set, which relies on studio interior sets whose Arabian flavour is achieved through oriental ornaments and curving architecture. On the other hand futuristic, simplified, massive urban structures abound in 3D computer games, partly because 3D editors allow easy creation of simple shapes, but struggle with complex structures. Both groups chose to model their stages outside the game-editor (using Autodesk's *Autocad* and Discreet's *3D Studio Max*) and import them into the game. The complex

Figure 1. Siobhan Fitzgerald's design sketch for one penthouse-like VE facing the skyline of a modern city for the *Casablanca* experiment.

geometries offered by these programs were ignored in favour of a cube-shaped, game-like ethos. This indicates a certain preconditioning of the participants towards the use of the chosen game platform, which was introduced to the participants in the beginning of the workshop. The aesthetics of the underlying game as presented beforehand influenced the spatial design beyond the technical limitations of the engine, producing an amalgamation of game aesthetics and dramatic expressive settings, as the groups mirrored the tension of the scene in the architecture of the virtual space. On the other hand, basic platform-independent expressive means were clearly implemented.

One group used height as an expression of the drama between the two characters throughout the scene, conveying the dramatic confrontation and changes of status they undergo through movements on stage (*mise en scene*). When Ilsa begs Rick to give her the transit visa, she stands on a lower level on the virtual stage, with the camera emphasising Rick's superior and powerful positioning. This contrasts with the *mise en scene* of the original movie scene, where both stay on an equal level throughout.

The second group divided the scene into three different zones: interior; threshold with connected exterior; and detached exterior roof patio—all part of a terrace-like structure. While the interior set expressed the introduction and build-up of the drama (establishing Ilsa's attempt and Rick's refusal), the threshold with connected exterior staged the point of highest tension (Ilsa demanding the documents from Rick at gunpoint), which leads to the turning point, where Rick literally turned around in space and entered the third area, a raised area of the terrace exterior, where the scene finally resolves.

Spatial structures such as the door towards the outside and the step up to the final area were implemented as dramatic 'landmarks' (Lynch 1960: landmarks in the cognitive understanding of space), and as such structured the space and the unfolding drama. Both groups produced virtual sets which instantiated the dramatic requirements of the scene as performance in virtual playspace, rather than depicting a *Casablancan* backdrop.

127

Figure 2. Status represented in the spatial definition on a virtual stage (3D engine: Epic's *Unreal Tournament*).

Figure 3. Interior setting at the beginning of the scene (3D engine: Epic's *Unreal Tournament*).

Figure 4. Final exterior setting at the end of the scene (3D engine: Epic's *Unreal Tournament*).

Use of colour

Casablanca was shot in black and white in a film-noir style, with highly contrasting dramatic lighting, still hard to reproduce in 3D game-engines. The virtual characters, avatar models of Ingrid Bergman and Humphrey Bogart, were textured in black and white and their overall appearance based on the original movie costumes. However, in texturing and lighting the scene, both groups constructed their virtual sets using dramatic colour-codes familiar from film and theatre productions. One group divided the scene-space into two main colour areas. Their scene develops from a glowing red during the first hostile confrontation, to a calmer blue at the resolution. In contrast, the other group staged the whole scene during daylight in pale light yellow shades, breaking the continuity of the original scene, set at night. They deliberately turned away from the gloomy design and claustrophobic interior settings favoured by computer games with limited 3D engines. Their decision owes more their reaction against that aesthetic than to the cinematic origins of the scene. Altogether, although designed with a cinematic eye to expressive narrative space, game aesthetics proved a significant element in content-creation, even for this non-commercial experiment.

Characters and their expression

The *Unreal Tournament* engine is designed for a first-person-shooter computer game. This means that the movements of characters are tuned to be fast, exaggerating leaps and other feats of physical skill, while instead of naturalistic spatial movements (like pointing, or sitting down) avatar animation design features shooting, being hit by virtual bullets, dying, and triumphant gestures and poses. This restricted pre-defined expressive range in RT3DVE will not be resolved through technical innovation in the near future, due to the available animation systems. Work such as Mika Tuomola's (Tuomola 1999) on incorporating dramatic gesture and characterisation inspired by rule-based Commedia del Arte, or by Hannes Vilhjalmsson at the Gesture and Narrative Language Group, MIT, may eventually be fruitful for a wider range of improvised performance in virtual environments. However, it is also possible to treat the gestural library currently available as a specific of the medium, and exploit it.

The *Casablanca* screenplay demands that Rick "walks over to a table and opens a cigarette box, but finds it empty" (Epstein, Epstein and Koch 1942 101). While 'walking over to a table' is a general movement operation in space, 'opening a cigarette box' is a specialised action with a unique object. In RT3DVE, for Rick to take out a single cigarette, light it, and smoke it would demand a suite of unique pre-fabricated animations hard to justify in terms of expense and effort. Such physical interactions between two avatars as touching, kissing, or any other non-violent or subtle character/character-interaction are not supported by off-the-peg shoot-'em up avatar animations, which are mostly limited to avoiding collision with other avatars when not acting aggressively, triumphantly, or dead.

In the *Casablanca* experiment, this library of actions did not adequately cover the dramatic needs of the scene, so the performers did not stick to the screenplay and its staging directions, but worked with the available avatar actions. Adapting to limitations and learning to control and use the avatars expressively was an important function of the rehearsal phase. The *Acting in VR* project at University College, London, demonstrated that performers can adapt to VR interfaces: "in spite of the relative complexity of the interface and the paucity of what was possible, the actors learned to use it, and mostly to use it well, in a remarkably short amount of time" (Slater et al. 2000 7). For the *Casablanca* experiment the limited range of customized gestures increased the significance of basic movements, and the workshop participants learned very quickly to utilise these. Turning round, stepping away, or shaking an avatar's head became important expressions, staged at relevant moments.

During rehearsals, another element in character-control and animation emerged. The models used for Bogart and Bergman (GGmale2 and GGfemale2, modelled and animated by Erik Gist and Greg Miller) include animations activated automatically from time to time independently of the player's command. That the computer program adds to a virtual actor's expressions, activating animations independently of the controller's intention, can lead to unexpected performances. "The text itself *plays*" (Barthes 1977 162) (Barthes' italics) with the player, constituting a contribution to the dramatic performance by the program.

During the *Casablanca* experiment, the interference of the characters' automated animations with the staging was a very visible aspect of the game-system's influence on the drama-performance. Most subtlety delivered through the performances of Bergman and Bogart was lost in the puppet-like behaviour of the acting avatars. Unlike the Hollywood stars, the performers in this experiment delivered their interpretation of the events like marionette players and in a certain distance to their expressive characters. The distance was even enlarged as the puppets delivered expressive movements beyond any user-control. Inappropriate gestures coded into the virtual actors' automated behaviour sometimes evoked comic reactions in the audience, undercutting the theatrical, dramatic staging of the scene itself, which echoed the melodrama of the original. GGfemale2's repetitive hand-gesture of hitching up her breasts when not in direct player-controlled action was far from appropriate to Ilsa's moments of stillness. Because defining the resultant genre either as melodrama or comedy becomes impossible, the audience's ability to follow the action and enjoy it is interfered with. This is an issue which needs to be tackled if RT3DVE drama is to be generated according to the conventions of one genre or another. Consciously applied, such automated characteristic gestures might positively create individuality in a cast member, just as they do in pantomime or dance.

Physical versus virtual camera

A comparatively high budget film such as *Casablanca* does not suffer from severe production limitations, but all limitations influence the aesthetics of the final movie for better or worse. For example, because of the ongoing war effort, no real plane was available for the final scene at the airport, so a plywood model was used. This had to be disguised by the massive fog effects (Harmetz 1993), which have now become the hallmark of *Casablanca*'s equivocal final sequence. A virtual production does not face these limitations, since virtual cameras do not depend on film-stock, and virtual sets do not need physical materials for their construction. Virtual cameras can depict any virtual spatial structure from any viewpoint.

In the *Casablanca* experiment, since no additional code was added to the *Unreal* engine, the camera work was defined through the positioning and orientations already available. The camera could not roll or tilt sideways but was otherwise spatially unrestricted, opening up a wide field of strategies not normally available to physical cinematography.

In this RT3DVE, the available viewpoints either depended on the positions of the visible characters (e.g. camera circling around a character at a constant distance, in the way invented by Alfred Hitchcock and Robert Burks for *Vertigo* (1958) and reinvented by Claude Lelouch for *Un homme et une femme* (1966) to express subjective feeling), or allowed free positioning within the VE—transcending Bazin's demand for realism (Bazin 1967) that asks for a camera that should be able to show the events in the whole of their occurrence and therefore from every possible angle—and enabling the creation of a different kind of moving-image narration, which gives the camera operator great interpretative freedom. In order to utilise this freedom to the best advantage, both groups laid out their camera strategies in storyboards before the actual sets were modelled, so that the virtual set could be adjusted to the visual concept. Once the virtual stage was assembled and rehearsing began, the use of the camera was refined through trial and error.

Since the camera operator constituted a third character on the virtual stage, capable of switching between different viewpoints, the final camera strategy depended on the personal choice of the operator, combined with the freedom of the camera in space. This made interesting demands on the process of set design. The original film-set, designed by Carl Jules Weyl, was itself 'larger than life'. In the

RT3DVE version, the participants exaggerated proportions even farther. The final performance developed from a ballet-like interaction between the characters acting and the camera mediating, on the expressionistic virtual stage, where both had to be able to move around freely. Although this performance was staged and followed its predefined visualisation and performance guidelines, it was not fixed as every live performance constituted a new version of the scene—for the acting characters as well as for the camera operator. The virtual stage was the only fixed entity during the performance itself. As such it had not only to fulfil its dramatic expressive function, but also to facilitate the practical functional task of providing the camera and the characters with appropriate access to the performance, just as in a traditional movie-set in a studio the ceilings and walls are designed to be removable, to allow the camera room to manoeuvre. The space of the drama in the VE proved not to equate with 'real space' or the architectural 3D fly-through, but with 'set design'—it had to be both expressive and appropriate for camera-mediated performance.

The camera—in contrast to the acting characters—was not bound to linear spatial movement during the performance. While the acting avatars moved through space in a coherent human way, the camera operator could 'teleport' freely. This virtual performance replaced filmic pre-planned continuity shooting/editing, which links individual shots together into a seamless sequence in post-production. Here, real-time 'vision mixing' in RT3DVE works more like TV football coverage or a live broadcast from a studio. The virtual camera operator edits on the fly within the timeline of the ongoing action, by 'teleporting' the camera to a new position to offer a new angle. Every cut is effectively a movement in space. The workshop demonstrated that in RT3DVE drama, editing, like the organisation of the narrative and the camera strategy, is a spatial function. Performing it well requires a proper understanding of spatiality as well as dramatic narrativity.

Theatre and cinema in the game

For the experiment, the virtual camera operators, used to 'real' video cameras to portray space and action, were quick to adapt to the features of the virtual camera available in *Unreal Tournament*. The evolving camera-styles of both groups soon abandoned the classical Hollywood approach of the movie *Casablanca* in favour of a fluid style more characteristic of the DOGME movement of the late 20th century. The number of cuts was small: each group used only 7 cuts in the scene, in comparison with the 25 cuts of the original; but camera movement was more elaborate. The operator positioned the camera as an active observer on stage—always relating to the other two characters. This was not the ideal observer totally mobile in space and time to which classical cinema has aspired, but an involved performer—a third character involved in the action, rather than a distant, all-knowing director's eye. The camera's behaviour, like that of the acting characters, was spatially defined.

Nevertheless, on various levels the final visual styles of the *Casablanca Interactive Performance Experiment* quote classic cinematic camera work. The virtual camera used foreground and background as depth-defining dramatic layers. It included reverse-angles, jump cuts, dollying, tracking, establishing shots, and close ups where dramatically appropriate. Here, the aesthetic influences of *Unreal*

Figure 5.
Floor plan of one
implemented VE
showing the pre-
planned spatial
movements of
characters and
envisioned
camera positions
(by Siobhan
Fitzgerald).

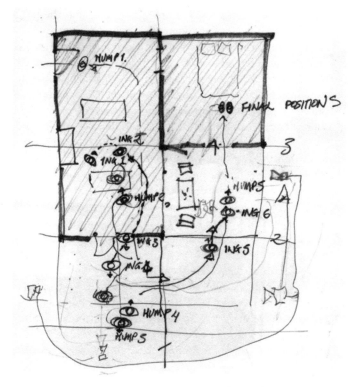

Tournament's gameplay which concentrates on a first person or tracking point of view—a camera perspective implemented briefly only once, with deliberate subjective dramatic effect, by one group—were minimal, and cinematic practice dominated. Although the creation-process of the visuals within the engine and the virtual environment was defined by game-functionality, a cinematic use of these functions was quickly established by the groups, whose aim was to create effective narrative drama.

Emergence of a *Spielform*

The *Casablanca Interactive Performance Experiment* demonstrated that participants with an educated sensibility in spatial design, familiar with game-engine language and traditional cinematic dramatic narrative media, were able to generate an engaging improvised 'movie' performance in a game-like RT3DVE. The improvisation extended from the visible acting characters to the camera operator and allowed for a complex interplay between all involved performers. The experiment proved that such interplay is possible in RT3DVE. In this early form the sound was deliberately excluded from any changes but future developments can include this layer (or parts of this layer such as the dialogue but not the musical soundtrack) in their performances.

The recorded results of both groups' experiments suggest that RT3DVEs can indeed create dramatic experiences for audiences as well as performers, using theatrical and cinematic techniques, while also exploiting the specific history, aesthetics, and functionality of the RT3DVE. Game, theatre, and film convey dramatic dynamic tension differently, each medium relying on its own visual techniques, functionality and mediation strategies. However, cross-fertilisation between them proved to create an unexpectedly immediate and engaging form of performance, which worked both for active participants and for spectators. It combined expertise (the original scenario and voice performances) with much less skilled performance-talent, but a reasonable degree of familiarity with both the games and cinematic creative media, to produce a genuine and effective experience.

A real-time event in a RT3DVE evolves from a mixture of cinematic mediation and theatrical performance. It demands specific spatial design that uses expressive space like the theatrical stage as well as the film set. The fundamental importance of spatiality delivers a crucial dramatic impetus during the performance itself in the co-operation between actors and camera operator on-stage, as well as in the dramatic shaping, colouring, and lighting of the stage/set. The navigability of the virtual playspace and the flexibility of the virtual set offer real dynamic engagement.

Preparing the environment for effective *mise en scene* offered the expressive set for the improvisation of the actor- and camera-performers in real-time allowing the maintenance of the tempo and credibility of the performance in this expressive narrative space. Space, performance, and visualisation combined offer a high level of involving dynamic interaction. This interaction, which evolves organically in the pre-production phase during the collaborative creative process between several active partners, delivers its reward in the heightened stakes of the actual RT3DVE performance. This creative process is not part of watching a movie in the cinema—where it has already happened during the production of the movie—but of the performing art of live theatre, where improvisation has long been a source of energy and involvement. The circle seems to be completing itself—film, theatre, and digital media converge to generate a dramatic *Spielform* for RT3DVEs that is not traditional theatre staged in cyberspace or interactive cinema for digital media or a typical narrative computer game. The *Casablanca Interactive Performance Experiment* shows that this emergent *Spielform* has great potential for worthwhile, engaging narrative drama in RT3DVE. It also demonstrates the value of RT3DVE as a platform for the development of dramatic techniques for interactive digital media. It proves that we, the human performers, can readily adapt to, and extend, the characteristics of this emerging platform.

Practical lessons learned

Using third-party modelling-tools increases flexibility for content-creation. Some 3D engines do not support import-features from professional 3D programs, and this can substantially restrict creative freedom. Spatial design for drama ideally needs to be free from such restrictions.

Using the recorded soundtrack from the film proved very helpful in the case of this four-day intensive workshop experiment. Limitation to a specified high-quality soundscape allowed participants to focus creative attention on the spatial organisation of the scene and the expressive potential of the space. The expertly dramatised soundtrack, with its evocative and carefully paced film music (by Max Steiner) helped the performers to time their *mise en scene*, performance and editing effectively.

During this workshop existing computer code, animations and the settings of the off-the-peg avatar characters were not changed in any way. Both characters included partly unsuitable animations. Adjusting the character avatar-animations to express the drama would be the first, most obviously interesting, experiment to try. This is not a daunting, but a time-consuming task. Adding customised code is equally possible, to extend the power of the medium by exploiting the potential of the computer itself to actively generate dramatic experience.

The creation of a team consisting of actors and a camera operator worked very smoothly both conceptually and technically, but future workshops might give the camera a visual presence in the RT3DVE. Because the camera was invisible during this first experiment, performers were not able to fully play their avatar characters with reference to the camera's mediation. Including a visible point of reference for the camera should enrich the collaboration between actors and camera operator.

A further experiment arising from the *Casablanca* studio workshop was made within the HEFCE-funded *Cuthbert-Hall Virtual College* (Haven) project (Nitsche and Roudavski 2002, 2003), where camera-control is partially shifted to a computer-program which runs a rule-based mediative system for actions performed within the RT3DVE. Other elements remain under the human interactor's control, and experiments to test how these combinations of interactivity and camera-control function to engage visitors in virtual place are under way at the Digital Studios.

The *Casablanca Interactive Performance Experiment* demonstrates that by combining spatial design and dramatic narrative in performance in the expressive space of a RT3DVE, a new and engaging *Spielform*, which both exploits and transcends tradition, integrating expertise and experiment, can be generated using extant games technology. The results of this practical experiment move the analyses and predictions of Aarseth, Murray and Manovich into the realm of actual performance-based activity in the computer-supported environment. This represents an important step towards theorising interactive narrative drama in an increasingly complete and satisfying way. Hopefully, it will also lead to the production of more engaging, rich and worthwhile interactive performance-based drama in the computer-supported environment.

Acknowledgements

This workshop would have been impossible without the help of Stanislav Roudavski. Thanks to the graduates of 2002 who performed the experiment at the Digital Studios, and especially to Siobhan Fitzgerald for permission to use her sketches for this article.

Note

1   The workshop was conducted as part of the MPhil degree in Architecture and the Moving Image, University of Cambridge, UK.

References

Books

Aarseth, E. J. (1997) *Cybertext: perspectives on ergodic literature.* The John Hopkins University Press, Baltimore and London.

Barthes, R. (1977) *Image-music-text.* Trans. Heath, S. Fontana, London.

Bazin, A. (1967, 1971) *What is cinema?* Vols. 1 and 2. Berkeley, Ca.

Benedikt, M. (ed.) (1991) *Cyberspace: first steps.* MIT Press, Cambridge, Massachusetts.

Drucker, S. M. (1994) *Intelligent camera control for graphical environments.* PhD thesis, MIT Boston.

Epstein, J., Epstein, P. and Koch, H. (1942) *Casablanca.* Unpublished copy of the original screenplay.

Harmetz, A. (1993) *Round up the usual suspects: the making of Casablanca.* Weidenfeld and Nicolson, London.

King, G. and Krzywinska, T. (eds.) (2002) *Screenplay: cinema/ videogames/ interfaces.* Wallflower Press, London and New York.

Laurel, B. (1993) *Computers as theatre.* 2nd edition. Addison-Wesley Publishing Company, Reading, Massachusetts.

Laurel, B. (1986) *Toward the design of a computer-based interactive fantasy system.* PhD thesis, Ohio State University.

Laurel, B. (ed.) (1990) *The art of human-computer interface design.* Addison-Wesley Publishing Company, Reading Massachusetts.

Ludlow, P. (ed.) (1996) High Noon on the electronic frontier: conceptual issues in cyberspace. MIT Press, Cambridge, Massachussettes.

Lynch, K. (1960) *The image of the city.* MIT Press, Cambridge, Massachusetts.

Manovich, L. (2001) *The language of new media.* MIT Press, Cambridge, Massachusetts.

Murray, J. H. (1997) *Hamlet on the holodeck. The future of narrative in cyberspace.* MIT Press, Cambridge, Massachusetts.

Poole, S. (2000) *Trigger happy. The inner life of videogames.* Fourth Estate, London.

Schrum, S. A. (1996) *Theatre in cyberspace. Issues of teaching, acting, and directing.* Peter Lang Publishing, Bern.

Stone, A. R. (1995) *The war of desire and technology at the close of the mechanical age.* MIT Press, Cambridge, Massachusetts.

Tomlinson, W. M. Jr. (1999) *Interactivity and emotion through cinematography.* Master thesis MIT Media Lab Boston.

Turkle, S. (1995) *Life on the screen: identity in the age of the Internet.* Touchstone, Simon & Schuster, New York.

Articles

Bolter, J. and Grusin, D. R. (1996) Remediation. *Configurations* 3 311–358.

Bolter, J. D. (1997) Digital media and cinematic point of view. *Telepolis* (14.02.1997) Hannover, Heinz Heise.

Cavazza, M., Charles, F., and Mead, S.J., (2001) Characters in search of an author: AI-based virtual storytelling. In *First International Conference on Virtual Storytelling*, Avignon, France, 27–28 September. Springer, pp. 145–154.

Craven, M. et al. (2001) Exploiting interactivity, influence, space and time to explore non-linear drama in virtual worlds. In *Proceedings of SIGCHI 2001*, Seattle, US, March 31–April 4. ACM Press, pp. 30–37.

Fuchs, M. and Eckermann, S. (2001) From 'first-person shooter' to multi-user knowledge spaces. In *1st Conference on Computational Semiotics for Games and New Media*. COSIGN *2001*. Amsterdam, NL, 10–12 Sept. pp. 83–88.

Hayes-Roth, B., Brownston, L. and van Gent, R. (1994) Multiagent collaboration in directed improvisation. *Stanford Knowledge Systems Laboratory Report* KSL-94-69.

He, L.-W., Cohen, M. F. and Salesin, D. H. (1996) The virtual cinematographer: a paradigm for automatic real-time camera control and directing. In *Proceedings of SIGGRAPH 96*, New Orleans, 4–9 August. ACM Press, pp. 217–224.

Lehane, S. (2001) UnrealCity. ILM creates artificial cities for artificial intelligence. *Film and Video* (7 July).

Mine, M. (2003) Towards virtual reality for the masses: 10 years of research at Disney's VR studio. In *Proceedings of the workshop on Virtual Environments 2003*, Zurich, Switzerland, 22–23 May 2003. ACM Press, pp. 11–17.

Nitsche, M., Roudavski, S., Penz, F. and Thomas, M. (2002) Building Cuthbert Hall Virtual College as a dramatically engaging environment. In *Proceedings of PDC 2002*, Malmö, Sweden, 23–25 June 2002. CPSR, pp. 386–390.

Nitsche, M., Roudavski, S., Thomas, M. and Penz, F. (2003) Drama and context in real-time virtual environments: use of pre-scripted events as a part of an interactive spatial mediation framework. *Proceedings of TIDSE 2003*, Darmstadt, GER, 24–26 March 2003. Fraunhofer IRB Verlag, pp. 296–310.

Pausch, R. T., Snoddy, J., Taylor, R., Watson, S. and Haseltine, E. (1996) Disney's Aladdin: first steps toward storytelling in virtual reality. In *Proceedings of SIGGRAPH 96*, New Orleans, 4–9 August. ACM Press, pp. 193–203.

Perlin, K. and Goldberg, A. (1996) Improv. A system for scripting interactive actors in virtual worlds. In *Proceedings of SIGGRAPH 96*, New Orleans, US, 4–9 August. ACM Press, pp. 205–216.

Saarinen, L. E. (2002) Imagined community and death. *Digital Creativity* 13(1) 53–61.

Slater, M., Howell, J., Steed, A., Pertaub D-P., Gaurau, M. and Springel S. (2000) Acting in virtual reality. *Proceedings of the Third International Conference on Collaborative virtual environments.* San Francisco, US, 10–12 September. ACM Press, pp 103–110.

Tomas, D. (1991) Old rituals for new spaces: rites de passage and William Gibson's cultural model of cyberspace. In Benedikt, M. (ed.) *Cyberspace: first steps.* MIT Press, Cambridge, Massachussettes, pp. 31–48.

Tuomola, M. (1999) Drama in the digital domain: commedia dell'Arte, characterisation, collaboration and computers. *Digital Creativity* 10(3) 167–179.

Films

Curtiz, M. (1942) *Casablanca.* Warner Bros., US. 102 minutes; black and white.

Hitchcock, A. (1958) *Vertigo.* Alfred J. Hitchcock Productions and Paramount Pictures, US. 128 minutes; Technicolor.

Jankel, A. with Morton R. (1993) *Super Marion Bros.* Allied Filmmakers, Cinergi Productions, Hollywood Pictures, Lightmotive and Nintendo Co. Ltd., US/UK. 104 minutes; Technicolor.

Lelouch, C. (1966) *Un homme et une femme.* Les Films 13, FR. 102 minutes; black and white/ Eastmancolor.

Lisberger, S. (1982) *Tron.* Lisberger, Kushner and Walt Disney Pictures, US. 96 minutes; Technicolor.

Sakaguchi, H. and Sakakibara, M. (2001) *Final Fantasy: The Spirits Within.* Chris Lee Prod. and Square Co. Ltd., JP. 106 minutes; DeLuxe.

Spielberg, S. (2001) *Artificial Intelligence: AI.* Amblin Entertainment, DreamWorks, Stanley Kubrick Productions and Warner Bros., US. 146 minutes ;Technicolor.

Wachowski, A. and Wachowski, L. (1999) *The Matrix.* Groucho II Film Partnership, Silver Pictures and Village Roadshow Productions, US. 134 minutes; Technicolour.

West, S. (2001) *Lara Croft: Tomb Raider.* Paramount Pictures et. al., US. 100 minutes; DeLuxe.

Digital media

Capcom (2002) *Resident Evil: Zero*. Survivial horror action game for the Nintendo Gamecube. Developed by Capcom. Published by Capcom Entertainment (original Japanese *Biohazard* series)

Digital Extremes and Epic MegaGames, Inc. (1998) *Unreal*. First person shooting action game on CD-ROM. Developed by Digital Extremes and Epic MegaGames, Inc. Published by GT Interactive.

Digital Extremes and Epic MegaGames, Inc. (2002) *Unreal Tournament 2003*. First person shooting action game on CD-ROM. Developed by GT Interactive. Published by Atari/ Infogrames.

Epic MegaGames, Inc. (1999) *Unreal Tournament*. First person multi-player shooting action game on CD-ROM. Developed by Epic MegaGames, Inc., Published by GT Interactive.

Konami (1998) *Metal Gear Solid*. Stealth strategy action game on CD-ROM. Created by Konami (designer Hideo Kojima). Published by Konami.

Remedy / 3D Realms (2001) *Max Payne*. Exploration and action game on CD-ROM. Created by Remedy / 3D Realms. Published by Gathering of Developers, Inc.

Team Soho (2003) *The Getaway*. Exploration and action game on CD-ROM for PlayStation 2. Published by SCEE.

Research Projects

Acting in VR (2000-) Slater, M. et al. University College London, UK. See also http:// www.cs.ucl.ac.uk/research/vr/Projects/Acting/

Avatar Farm (2000) Benford, S., Greenhalgh, C., Craven, M., Taylor, I., Purbrick, J. and Drozd, A. CRG, University of Nottingham, UK. See also http://www.crg.cs.nott.ac.uk/ events/avatarfarm/

Aspen Movie Map (1978) Lippman, A. MIT, US.

Haven-Cuthbert Hall (2001–) Nitsche, M. and Roudavski, S. Digital Studio/CARET University of Cambridge, UK. See also http://www.caret.cam.ac.uk/hefce/haven.htm

Out of this World (1998) Benford, S., Greenhalgh, S., Craven, M., Morphett, J. and Wyver, J. University of Nottingham, UK. See also http://www.crg.cs.nott.ac.uk/events/ootw/

Romeo and Juliet in Hades (1998–99) Naoko, T. and Nakatsy, R. ATR Media Integration and Communication Research Labs, Kyoto JP. See also http://www.mic.atr.co.jp/~tosa/it/ index.html

Virtual Reality Notre Dame Project (1998–1999) DeLeon, V. et al. Digitalo Studios, USA. See also http://www.vrndproject.com

(All links accessed March 2003)

11 Sorcerers' Apprentice: reactive digital forms of body and cloth in performance

Jane Harris and Bernard Walsh

Jane Harris' work reflects the evocative nature of material and is informed by contemporary and historical textiles. The diverse characteristics of cloth provide an intrigue that has fundamentally directed a collection of work, which is now entirely 3D computer graphically (CG) animated[1]. They reference former physical process and practice both in terms of the textile/garment form, the choreographic body movement and the invisible body image, absence and presence (Figure 1). The folds and shadows in the 'fabrics' reference the traditions and representations of cloth in the historical context of fine art and design. The visual reference to the body in these animated works deliberately renders the garment as a cast, emphasising the beguiling qualities of fluid material and form. The choreographed and layered movement in the pieces explores space and time. The works speak not only to the histories of fabric depiction but also to the contemporary context of new technology, contemporary dance, absence and presence. The work transcends category and serves to inform us of an alternative world that is digitally constructed, one that looks beyond simulation of what we know and creatively

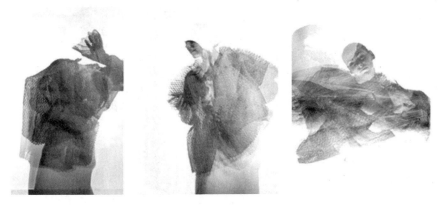

Figure 1. *Through You*. Costume/performance, Jane Harris, 1994. Photography Shannon Tofts

challenges our perception of what is 'real'.

Subsequently, through this use of digital media the development of performance and movement concepts that otherwise could not be explored physically, has evolved. Overcoming initial technical limitations of digital imaging media, particularly early on in 1995/96, offered a new perspective on working within a CG constructed 3D space and so defined different methods of practice.

The additional component of a 3D CG cloth/garment form forced consideration of unexpected behavioural elements, which now determine the thinking behind the choreography of the body and movement narrative on many levels. Optical motion capture is used to digitise the choreography of the human form. The aesthetic nature of the piece is composed of numerous parts, which function as a whole. In this sense a still image does not convey the complex language of movement between body, garment and other reactive elements within the 3D space.

In terms of evolving practice, process and the nature of collaboration, this chapter will explore issues that have arisen through the development of this work: how the choreography of body and dress evolve; how these two elements function as reactive forms and present particularly evocative imagery and experience; pursuit of identity within these pieces; sound concepts which will ultimately accompany future work; the notion of this work evolving into real-time and what that means in terms of live performance; the nature of the audience that the 3D computer graphic animation work draws.

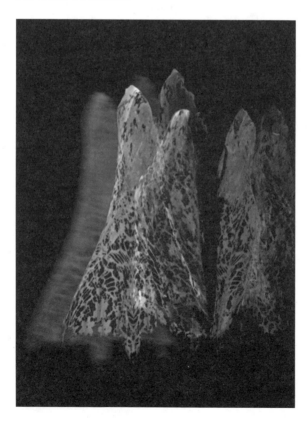

Figure 2. *Portrayal*. Jane Harris, 2000. Movement Ruth Gibson. CG work Mike Dawson.

Bernard Walsh is an artist and writer. His work incorporates practice and theory, which plays visual imagery into written/presented texts. Focusing on 'place', and in particular definitions of the 'commonplace', his work challenges the authority of perceived values and conventional standards, interrogating the status of 'high' and 'low' art in relation to the continuous development of contemporary art practice. His experience of digital imaging technology, however, is limited and he therefore questions this work from a lay perspective in relation to the processes involved. As the viewer is an intrinsic part of the work, this paper has been developed as a dialogue between Jane Harris and Bernard Walsh. The following is an outline of their conversation.

*

BW: I viewed an early piece of your work, *Portrayal* at *Pixel Raiders*², March 2002, as part of a series of digital installation/performance pieces. The piece worked for me on several levels. Suddenly I was spellbound by what I saw as an elegant form, a dress, appeared to come from nowhere and drift across an otherwise empty stage. The body inside was invisible. However the function of the body form remained a presence as the movement suggested a pivotal motion, the body form inside. Then I wondered about this presence. How did it operate? Where were the usual means, physical support or visible signs? I continued to watch a sequence of courtly gestures beckoning, exciting, alluring me and forgot to think about what was practical, as the phenomenon already existed.

The work is so 'artfully' constructed.

JH: 'Artful' is such a conscious word.

BW: Perhaps I mean that I cannot figure out the process behind the work.

JH: Good! Because, I don't want the process to show. But by the same token the work wouldn't be possible without all of the processes involved. It is necessary to use these particular methods because it achieves what would otherwise be physically impossible. I don't think that the process is immediately important for the viewer. It is an additional element, that might be interesting in a discussion about the work.

BW: Why is that important?

JH: Because I want the viewer to enter into the space and find the work there through their own sense. The form of the presentation provides a framework, a surreal parameter through which the viewer might wonder. Which is why everything has to seem as though it is real and the process has to 'appear' invisible. But I recognise how essential the process is.

Working digitally is strange because everything is divided into separate stages as a process. When you work physically you don't have to think so consciously about the process in that making because of the intrinsic processes involved, accepted processes, almost subliminal processes, the experience is more fluid. Whereas working digitally you have to think about the processes involved in a much more conscious way. The range of participants and layered use of technology require specific planning in the execution of the work. This may seem to be a hindrance and a contrast to physical working practice. However as with other limitations in working with digital media, knowledge of those limitations is evolving into new forms of practice.

For example, early on cloth animation would malfunction if the movement driving it; that is, the choreography was too fast. This forced enquiry of movement, so slow as to be unfeasible on a physical stage with the costume that we were constructing at that time.

Another example is collision detection, which in CG terms describes the recognition and response of one object or 3D form to another. For example, when cloth or garment hits a computer graphic body and responds to the existence of that body. Without collision detection, the cloth visually appears to fall through the body at certain points of contact. One of the most technically complex CG object recognition issues is cloth recognising cloth, which is a problem if a garment has layers of material or lining for example. Having resolved various related technical issues and as software has improved, collision detection has evolved as a useful tool that we are in the early stages of consciously incorporating into the choreography to unusual and intriguing effect. These may seem like basic considerations or simply gritty elements of process but this level of required consideration has immediately changed all parameters of thinking with regard to working in this sort of space.

BW: Do you mean a virtual space?

JH: Virtual is the wrong term because it implies mimicking reality and I think that I am playing with a sense of reality within the digital space in a way that I can't in a physical space which is a different kind of 'stage'. I want the viewer to be convinced by what they see, to register a reality, to perceive a total depth.

BW: Then you are challenging the viewer to enter into a different realm?

JH: Yes but not in a confrontational way, I am trying to draw them into the work, to be beguiled by it.

BW: Through an artful device?

JH: Yes! because when the technology works, when it's invisible, it can be a very seductive tool.

BW: When deploying this strategy do you have a specific audience in mind?

JH: The audience is extremely important. I feel the frustration of the audiences that turn away from works made using digital technology, particularly interactive works that don't truly connect with the viewer or reach expectation. I think the viewer's ability to interact with a really good visual, such as dance or performance can produce, is totally underestimated. This contrasts hugely with some interactive works, which often leave an audience cold, lacking an understanding of the piece and even alienation or exclusion from the process, which can be shrouded in complexity as an explanation. My work is visually extremely simple and a specific aesthetic balance is achieved, which attracts a surprisingly broad audience. The pieces so far are silent and yet children sit in front of a five minute loop of the animation for half an hour. They don't even realise that it is computer generated and even if they do, there are none of the draws of computer generated work such as they know from particularly the games sector i.e. sensory tools of speed, noise and garish special effects. Equally the costume is a very present element capable of drawing a male audience, this would not usually be the case of a textile exhibition, which draws a substantially female audience.

A short pause before Jane laughs and then quietly explains her general observation.

JH: I am thinking about a woman walking down the street, her dress fluttering in a light summer breeze. Bob the builder is working over-head, watching her tits and body moving inside the dress. Or is it the dress that is actually evoking his stare? If the garment was suddenly walking down the street alone, without the body inside, would he watch it with as much intrigue? I suspect that he would but would never admit that it is the dress that is actually steering his gaze. Similar and enduring in its appeal is the Marilyn Monroe[3] image of the dress being blown up in the air, performing around her body as she attempts to hold a section of the skirt down around her legs. The performance is twofold, a sheath that responds unpredictably to external elements of wind and air and the body within, that is contained by this sheath and yet also has influence on the overall movement effect. There is something about the evocative and the unpredictable characteristics of cloth that intrigues me. The quality of a material will only become self-evident as we begin to touch. But before then, before that touch, there is only the suggestion of that quality.

BW: How we are able to form an association between specific materials and particular qualities, is that what intrigues you?

JH: Yes but the material needs a realistic shape staged through the performance of movement of body and form, this is a crucial kinetic component. Within a CG environment it is possible to have an element of control over the costume for example, which may also be used to specific effect, this would not be possible in a physical environment. Often in contemporary dance the visual emphasis either lies purely in the body and its form interacting with other forms or the use of costume, which overlies the body.

BW: But the body isn't real in your work.

JH: Yes it is. The body form is rendered invisible within the costume but it is present through the suggestion of the human movement, which is real. In fact more real in a sense. The digital 'skeleton' of data achieved by the motion capture (MC) process is exact. The technology was developed primarily for the medical industry to measure the gait of the body and the movement characteristics that are captured are very true in every sense[4]. This is one of the reasons I am interested in specifically exploring a range of characteristics, using this technology in relation to the gestural nature of the body in a narrative form.

Presently the visual or narrative concepts I seek to convey are discussed and explored with choreographer Ruth Gibson[5]. I try not to limit the choreographic ideas that Ruth creates in relation to the overall work, a process

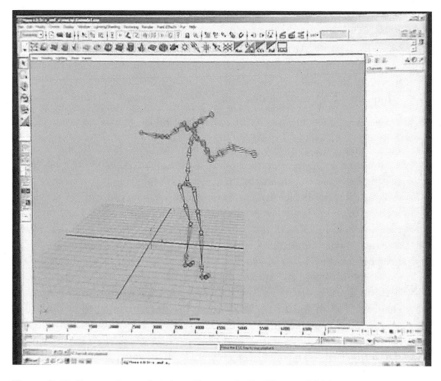

Figure 3. Motion capture skeleton. Dancer Ruth Gibson, 2001.

in itself, which has evolved over the time we have worked together. But there are points where intervention is appropriate and a type of collective digital and physical tacit knowledge informs this, based on both physical and digital cloth characteristics and their reactionary properties in accordance with the body's motion as a guide. Ruth has a movement vocabulary that she has developed through her own work, which she draws upon but there are specific movement patterns that are developed for the 3D CG work.

BW: There are so many variable factors at play here, notions of physical reality, surreality, virtual or digital forms and spaces. This implies some contrivance and yet I know from viewing the work that what I find is an accessible experience that overrides my suspicion of the potential gimmickry that digital tooling provides.

JH: I am very aware of the draw that digital technology has on a funding level as well as the fascination for its potential in a live context but I feel that a great deal has yet to be achieved with tooling that has already evolved, the use of which requires fine tuning aesthetically so that the experience of it over-rides the implied potential. My role as the coordinator is to make sure that our efforts are effectively combined so that the whole work makes sense. I am going back to the craft on several levels and that includes the 'making' of the visual components, the 'making' of the choreography with Ruth Gibson and the 'making' of the graphic animation with Mike Dawson. This is not to say

Figure 4. Working in the studio, Jane Harris and Ruth Gibson, 2001.

that I don't see the work functioning in a live context but the technology has a long way to go to enable this without compromising what I am presently pursuing aesthetically and conceptually. The pieces proffer and communicate the notion of performance, even though at present they solely take the form of installation.

BW: But the sense that you are trying to convey, what is that?

JH: Perhaps it begins through my own sense, or my own 'nonsense'. My original intention was to make what might have seemed impossible possible. Ruth is a fascinating dancer and choreographer, who uses her intelligence and physical prowess to extend the limitations of her practice. I have an experience of working with fabric. I know how to make material come to life. Mike is a computer animator who actually executes the work technically. He transforms the medium that we are working with through the development of the work that we are making together. Even if we sometimes don't always seem to be speaking the same language we are always learning from one another and sharing our ideas.

BW: So your individual skills combine within the work, as the work provides you with a common sense of purpose. But because of the processes that you go through and in terms of what you want to achieve as an overall effect, much of that effort is hidden. As you say earlier "the process has to appear 'invisible'. " Isn't that a contradiction, how can a "process appear invisible"?

JH: There are issues that arise out of the process that we have to confront, as it becomes a part of the work. For example the invisibility of, and we still have to do it, the invisibility of certain objects within the space which might function like the body does at present, as a reactive force. But what we are working with at the moment is just the invisibility of the body, as the movement of the body is essential as we play with the kinetics of the cloth. Everything else within the space has to be invisible to place the emphasis upon the movement of the cloth within the garment form.

BW: So what is the next stage in the development of this work?

JH: Knowing the body will appear invisible, although visible, through the cloth/ garment form, is an image I have to work with in my mind in terms of assisting in directing Ruth's movement. I also have to work with a mental image of how the 3D CG garment will function with the body data installed. Some body movements generate particularly effective or evocative cloth movement, affecting the total behaviour of the material as it reverberates over and around the body, generating a secondary kinetic and responsive movement and skin as a part of the overall combined kinaesthetic affect.

A more predictable approach to this working process would be to view how the garment responds to Ruth's movements during the motion capture, in real-time on a monitor prior to the lengthy CG process. Such a concept isn't quite possible yet, although inevitable, however this may relinquish the unexpected that informs new visual and creative vocabularies during the development of the CG animated installation pieces. Presently I tend to envisage a CG garment responding to Ruth as she works in a physical studio space. This may be like imagining sound, in relation to motion, that doesn't yet exist. The design of the garment for the installation pieces to date has been a visually simple, flowing, full length, sleeved form, which functions as a cast that responds to all the nuances of the body's movement from top to bottom. The design of the CG garment is however becoming more CG tailored and will have structural characteristics that will be effected by some actions. A balance of body to cloth movement in the final work without either gimmickry of the technology being evident or the dynamics of the cloth/garment overriding the choreography, has been an important objective in the overall work.

BW: The dis-'appearance' of the body reminds me of something. As I am looking into the work I am forced to go beyond the apparent frame, the medium informing the work, to explore the content. This content is not only what is present, presently presentable, but also what is not. Form and content do not necessarily relay the same sense. Content is beyond the form as a passage of inter-action takes place between the viewer and the specific medium. This is where the quality of the work is assessed, where it comes to life, as a sense of the subject is revealed through the absence of the body.

JH: Absence is notable because it remarks on something otherwise not present and proffers a beguiling element in creative work whether it be performance, visual or applied art. The play on illusion arouses curiosity, a questioning of

Figure 5. *Potential Beauty*. Jane Harris, 2002.

what we know and don't yet know. The invisible body in the installation pieces ironically draws even more attention to the form of the body. Somehow the movement of the body within the garment, appears to function quite powerfully in its invisible state. In an apparently floating world, an evolving theme in the work will be the body/garment response to other objects or surfaces in the CG space, rendered invisible.

So far the first invisible element included in the work, other than the body, is a floor, which isn't immediately apparent. Sporadically it is evident as the garment hits the surface of the floor and responds. Experimentation with other forms will include those that offer different platforms or spaces for the choreography to take place.

As contact is made with another invisible object, an additional force on the CG skeleton will be apparent through the cloth's reaction. The garment is the informant as to what exists in the CG environment and how the body is interacting. For example, dress brushing past invisible object or another body, in the environment. The cloth reveals the form. Through this a whole range of interesting and unexpected pressure points may begin to evolve. If the body or objects were visible, the cloth behaviour described would go almost unnoticed as perception of consequential body and garment responses are defined by prior knowledge. Abstract and perhaps impossible forms of movement could emerge as the focus, a play on visual kinetics then arise in the choreography. Whilst technically complex in CG terms, the effect appears visually simple although open to question.

BW: This is interesting because another effect of the work, brought about by the removal of the details of the original body, the first portrayal, is that the viewer is able to transpose a sense of him/herself. The work becomes interactive, inclusive of us, as we are able to imagine ourselves within the space, even though that situation is physically impossible. A sense of distance between us disappears.

JH: I think I allude to this earlier on by referencing how much the audience is underestimated in its interaction of both moving and static works of performance or art, even though they may seem to be static or are static because of the position that they are put in as viewers. As the animated body form moves, I have observed various limbs of the viewer mirroring this.

BW: I did feel myself becoming more graceful.

JH: Take this notion further and at some stage, technology allowing, the audience, if I wished, could be more directly involved, using the motion capture (MC) system.

BW: As a link between the pre-recorded patterns of another person's presence and our own individual and transitional sense.

JH: Yes. As a means of digitally representing the exact movement characteristics of the human form, MC technology can be used as a tool to explore concepts linked to identity and portrayal. I have previously considered this in the context of an invisible human form, e.g. myself, in the MC space with Ruth. Ergonomically Ruth and I are very different in scale and these characteristics are apparent in resulting experiments.

MC data for film and games in a commercial context is often created with the help of actors. The most realistic data is achieved when the subject is unaware that they are being 'captured', as the movement is particularly natural.

BW: But I am not naturally graceful.

JH: Bernard, in this context I am more interested in the potential.

Linked to the concept of invisibility a quite secretive form of portrayal or portrait will be explored, working from details of the individual. I hope a conversational narrative will evolve.

BW: As there is no longer a physical stage, entrance/exit sign or performer in place, instead there is an interesting threshold, a space where potentially anyone can perform.

But in order to evolve a conversational narrative there would have to be an unpredictable element at play throughout the process, as a method of exchange. How do you imagine this relationship occurring?

JH: Working live with optical MC has been possible for some time. Technically linking the MC to the garment in real-time would require sophisticated rendering tools. Another generation of games consoles and processing power have yet to be brought to the market in order to enable prolific use of 3D CG cloth[6] in a live context. Presently the aesthetic potential of this work would be limited in terms of movement and overall appearance. Although not impossible in terms of the available technology, there are first generation real-time cloth software tools available, the concept has not been extensively explored in other fields primarily due to cost and the difficulty of achieving a reasonable cloth aesthetic. The real-time representation of cloth has been generally quite crude, rubbery or synthetic in appearance, which fits the image of the games industry but is not for example appealing to the fashion or textile arena that could consider developing this technology. To maintain the detail and quality already achieved in current animated work will be very challenging. Complex garment structure, surface decoration, transparent effects or any additional or superfluous detail applied to the CG fabric adds considerably to the rendering time, hindering the objective of live performance work[7].

A purpose for working in real-time would be to introduce unpredictable elements.

BW: What, like my lack of grace?

JH: That is not necessarily unpredictable.

A combined use of choreographed and improvised work would drive the 3D CG body and garment form. Sound will become an important interactive element in this work, which up to now has not been linked to the animation work. The notion of CG costume could also be further developed experimentally, enabling movement/dance in costume that would otherwise not be physically possible to work with in a live context. As was the case early on in the CG animation work, it is also possible that the technology will not work in a manner that is expected, which may present another set of tools from which new working methods and thinking regarding the choreography in particular may evolve.

BW: At present the silence is interesting in itself and seems not to have deterred audiences.

JH: No

BW: As the sound informs a separate sense, maybe it should not be attached to the original composition. If placed beside, or played away from the visual work, an accompanying sound might emphasise qualities that you have already achieved within the visual performance. And I am thinking most particularly about the illusion of depth. Sound might echo that quality.

JH: Those are interesting ideas. I want to consider sound, initially working from physical cloth and body as a source the aim is to devise sound, which can be as subtle and evocative as the motion generated by physical interaction of these core elements. The present work belies material origin and CG process, which I think adds to the intrigue of the illusion.

BW: So the sound would have to realise a similar aesthetic balance?

JH: If the sound remained absent it would add another complexity to the illusion. The aesthetic balance of the work is based upon my own judgement. I try to anticipate the audience's response. But I am not trying to create a 'shock of the new'. In many ways the work that I am presently producing relies upon my earlier experiences working with the physical manipulation of material. Without a craft-based knowledge of cloth the aesthetic balance of the work would be very different.

BW: Does that mean that your audience has changed over the years?

JH: When I used to produce one-off garment forms my work was exhibited mainly in a textile environment. As the work evolves I gain knowledge. Working directly with Ruth and Mike and even this conversation with you alters my perspective. I am continuously re-evaluating what is aesthetically and conceptually possible. I am interested in the work reaching out to a wide audience, which is why I want it to be shown in a number of different contexts. I don't want it to become pigeon-holed.

BW: There is something about the uniqueness of the work you produce that brings together a range of different processes, which I understand is why you don't want the work to be categorised. But in order for the work to be critically evaluated it may need to be referred to beyond its technical accomplishment.

JH: I know that the work potentially straddles many spheres, that knowledge is where I begin to think about making work but the experience of actually making work, is a process that makes a lot more sense to me.

BW: And the viewer completes their own sense, because in my opinion that is what defines art, an avoidance of prescription. Then the art is within the speculation of the narrative, which exceeds the limitations of the form.

JH: I think that is true, as a consequence of what might exchange. But so much has changed, and is always changing, not least of all the technology and access to that technology. I don't think that there is a fixed concept of 'art' or a fixed concept of the 'audience' for art. As the advent of photography, in the mid-nineteenth century, threw into disarray previous assumptions around the nature and notion of representative forms, so the introduction of digital technology is having a similar effect.

BW: So you wouldn't mind Bob the builder getting interactive with your work?

JH: I think the notion of gender exchange is an interesting issue that we have tentatively explored but only by accident. Mike recently took some of Ruth's movement data and placed it inside a male CG wire mesh form just to view the movement. Offering another sample of the potential for the digital medium to distort or play with what we wouldn't otherwise know or recognise. The notion of the expression of Bob the builder or yourself driving a female body form within the costume is a challenging visual concept, another irresistible play. The consequence of which may be too subtle for the audience to translate. But again this thinking refers back to the audience's relationship with the process and their role in simply addressing the visual or the consequence of our play, which potentially offers an ongoing dialogue through the form.

BW: But there is also something sexy about the opportunity to be become another persons physical shape. Often it is the difference between our shapes that begins the attraction. The chance to impersonate or to emulate another person's shape could let loose a range of playful possibilities.

JH: I think the notion of identity through dress is an interesting field, the notion of guise of disguise through costume is something an audience seldom has the option to actually explore but in practice is potentially a liberating and revealing concept. Take this into performance, invisible performance, fantasy

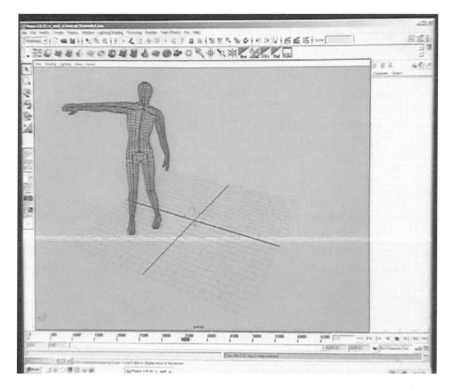

Figure 6. Ruth's Movement data within a male wireframe form, 2001.

and identity open up to the tools we are exploring digitally. The fact that these tools offer the space for invisibility suggests a place for this kind of exploration that is more subtle than the reality TV type concepts that on a more brash level presents the audience as both subject and viewer interactively.

BW:	So like the wardrobe in Narnia there is the potential to walk into other worlds. As the portrait remains one of the most popular, enduring and beguiling forms of visual representation the prospect of using CG animation to elaborate the theme of 'ourselves' being represented is exciting. Particularly if we are able to mingle the sense of 'ourselves' with a sense of other selves within the form of a moving presentation. This makes me feel as though there is hope even for me, and for others like me, who missed out on ballet classes as a child. This may seem absurd, but the audience for that work is potentially massive, particularly if we are able to perform. The popularity of reality TV proves this. Night after night so many of us tune in to watch 'ourselves'. But reality TV is, of course only a dreamworld, where 'reality' unfolds as a version of reality. As more and more attachments disappear the plot thickens and we are drawn in. With eviction, the show begins.

Credits

Paul Hallam
Ruth Gibson
Mike Dawson
Central Saint Martins
Goldsmiths College
Arts & Humanities Research Board (AHRB)

Notes

[1]   Ph.D. study at the Royal College of Art (RCA) (1995-2000) established the use of relevant digital media and collaborative partnerships, with individuals and industry that enable present work.

[2]   www.pixelraiders.org

[3]   *The Seven Year Itch*, co produced by Billy Wilder, 1955. A famous shot of her standing over a subway grating, with the skirt blowing up to her face.

[4]   Motion capture is a technology that generates human movement data and can be used to drive the movement of computer graphic characters in CG film and games.

[5]   www.Igloo.org

[6]   This comment is linked to the games industry as it is possibly the market that has invested most in CG software and hardware development.

[7]   An AHRB Innovation Award is presently supporting enquiry of a real-time cloth and motion capture enquiry linked to a digital dressing up concept for the museum environment. For this project we are working with a detailed 18th Century garment from the Museum of London collection, a particularly challenging piece. The technical outcome of this research will assist further enquiry of real-time use in the context of live presentation or performance in the future.

12 Turning conventional theatre inside out: democratising the audience relationship

Christie Carson

The digital revolution has not only changed our methods of communication, it has changed our vision of what kinds of communication are desirable. Mass production, the great success story of the industrial era, has been superseded in the post-industrial era by niche marketing and boutique shopping. The wonder of the digital world is the ability to create in bulk but to customise that product to make it appear individual, personal and original. Quite a lot of attention has been paid to the new kinds of performance that new technologies allow, however, I am more interested in the impact of new technologies on traditional theatre practices. If theatre is to continue to act as a mirror to society it must engage with the changing means of communication which new technologies have brought about. Digital technology offers the institutional theatres a real opportunity to reinvigorate and re-legitimise themselves as centres of public debate[1]. I suggest, however, that real responsibility accompanies the expanded remit which an increasingly democratic communication entails.

Creating a new audience relationship: a new opportunity

The advent of the Internet provides an opportunity for the large institutional theatres in Britain to redefine their relationship with their audiences. In this chapter I would like to address the contrasting approaches taken by the National Theatre, The Royal Shakespeare Company and Shakespeare's Globe Theatre in terms of making their work more public and more accessible through new technologies. In each case these theatres are taking on a more expansive role moving into areas of education and archiving, contextualising their current working practices and creating their own histories. I would argue that the results are of varying quality. The web presence of these theatres, in particular, exposes the strengths, the weaknesses but perhaps most importantly the biases and priorities of each approach.

The objective of this chapter is to assess the potential developments in this area and to offer possible models for future development. The economic pressures placed on these theatres have forced them, in the first instance, to follow business models for online audience development. I would argue that it is essential to separate the use of new technologies from business practices during this embryonic period in order to be able to discover the full range of new forms of creativity and communication that are possible through digital processes. Research and development in this area are increasingly out of the reach of individual institutions. However, I suggest that through collaboration the institutional theatres can participate in an experimental approach to new technologies which will provide the public with a range of options for the future. I also suggest that if the theatres move on their own into the domains that have formerly been the preserve of the libraries and the educational establishments they will open themselves to the criticisms that have been levelled at the BBC which has taken on an ever expanding role in public life through digital technology. The choices made by the institutional theatres in this area may well have a large impact on how theatre is taught and how it is perceived in the popular imagination. I put forward the suggestion that by working collaboratively the institutional theatres and educational establishments could help to create new models for the study of theatre. Used creatively, new technologies could offer real lines of communication between audience members and publicly funded institutions in the UK. The question which remains is, will the publicly funded theatres take seriously both the opportunities and the responsibilities offered to them through this technology?

The evolving influence of new technology in the theatre

Theatre is primarily about a momentary relationship between an audience and the actors on stage. The large institutional theatres have an ongoing responsibility to not only please current audiences but to develop new ones. This tension, particularly in a rapidly changing world, puts great pressure on the artistic staff of institutional theatres. Technology has increased this tension. The theatre has been struggling for some time to find the most appropriate role for technology in the live ephemeral event. By looking briefly at the response of the theatre to new technology over the past decade it is possible to illustrate the awkward nature of this relationship and to understand more fully the current challenges that face these theatres.

While digital technology has made its presence felt backstage for some time it was not until the 1990s that this technology began making an appearance on stage as well as behind the scenes. The use of mobile phones and computers on stage as a reflection of our contemporary society was an inevitable step in the integration of new technologies in the theatre world. Interestingly, the creative presentation of computers on stage raised new technical hurdles which caused all sorts of problems which had to be solved by the technicians. An example of this is Patrick Marber's play *Closer*, which was performed at the National Theatre and involves a scene in which two characters correspond through the computer. This scene brings together the two uses of digital technology in the theatre, as a technical tool and as a creative tool, in an interesting way. The two actors are silent throughout the scene but it is essential that the audience can see what they are typing. The dramatic irony of the scene lies in the fact that both characters are male but one is pretending through the computer chat room environment to be female.

This scene provides both a series of challenges technically and a series of very real acting and staging challenges. How can the actors be linked with the words they are typing since it is essential to the scene that this link is maintained? On a more practical level how it is possible to generate an accurately typed series of lines that appear to be spontaneous night after night? The solution to these problems was in fact linked in performance. The correspondence between the two characters was displayed on a screen at the back of the stage. This shared video screen was also seen by the actors on their respective computers. The actor, at the beginning of each line pressed a function key on his keyboard which set off the command for the video projection to type the pre-recorded line. This line was visible to both actors as well as to the audience so that the typing that the actor continued to do on his keyboard could stop at the appropriate moment. So the seemingly real time exchange of rapid typing was, in fact, an elaborate pre-recorded technical trick. It is interesting to see how in the theatre the attempt to illustrate the spontaneous nature of digital communication caused so many difficulties in execution in the same way night after night. In a way this example points out very nicely the advantages and disadvantages of digital technology in the theatre. Digital technology is entirely consistent and repeatable, human action is not.

The large stage productions of the Disney corporation offer another useful example in that they show an interesting learning curve in the relationship between digital technology and its use on stage. With *Beauty and the Beast*, the corporation's first foray into live theatre, no expense was spared to create an automated performance that could be replicated the world over. Not only was the show built for franchise, it was also built to mimic almost entirely the animated version of the story. Disney audiences, it was assumed, would come to the theatre expecting to see what they saw on their television and movie screens. In this case digital technology was used to power the onstage action and to keep the actors in line with the roller coaster ride of the performance. While successful with many children this production unnerved an adult audience. Richard Green writes of the touring version of the show:

> *It's all aimed at extending the mythology of Disney into our own lives, but as with most bus-and-truck productions, the artistic latitude in this reincarnation is so limited that I began to feel once again that corporate contrived sentiment is rather a blunt instrument indeed.*
> (Green 2003)

It is interesting that in their second experiment in live theatre a different approach was taken. *The Lion King*, which has had a much longer run than its predecessor, did not try to replicate the animated version of the story. Instead Disney, quite surprisingly, hired an experimental theatre director and gave her a fairly free hand in her interpretation of the story. As a result, the actor's bodies and individuality were emphasised rather than hidden beneath automated costume pieces. The use of masks, that extended the actor's bodies but did not hide them, proved much more successful on stage in the long run. Therefore it can be seen that there has continued to be a natural resistance in the theatre to the imposition of digital technology as part of the creative process in a way which is restrictive and which tries to hide the visceral human elements of liveness. The fallible nature of the live event has continued to dominate over the rigors of the technical program. Experiments like those mentioned above proved that while technical processes can be automated and on stage action can be enhanced by creative use of technology, both in content and in form, it cannot be replaced by that technology. The fundamental advantages and disadvantages of the different ways of working have, then, become increasingly apparent in the theatre. Digital technology encourages spontaneity and experimentation yet it is infinitely repeatable without loss of quality. Traditional theatre practices require planning and rehearsal yet they are unpredictable and impossible to duplicate, but therein lies the attraction.

Current shifts in working practice in the theatre

I suggest that we are starting to see a shift in the nature of the ongoing relationship between the National and the Royal Shakespeare Company and their publics. In the first instance it was marketing departments and development departments that were responsible for the outward face of these theatre companies. It was these departments that made contact with the audience outside the theatre and they determined the nature of that relationship. The exchange, ticket revenue or donations in return for the theatrical experience, prestige and perhaps even a lifestyle, was based primarily on commercial principles. The shift that seems to be taking place in the new century is a move towards an audience-driven relationship which is shifting the responsibility of the content of that interaction from marketing to education and to an interaction with the theatre artists themselves. Through web-based archives, projects and interactions the institutional theatres are moving towards creating an ongoing relationship with their audiences which is based on an interest in and an engagement with the theatrical process. These new forms of interaction are allowing for a discussion with audiences that can begin long before the audience arrives at the theatre and can carry on long after the experience of the live event is over.

Not only are shifts apparent in the audience relationship outside the theatre but both the National and the RSC have started to make changes to their repertoire which facilitate a more personalised experience. The size and complex bureaucratic nature of large theatres work against major shifts in working practice, yet the need to develop new audiences means that changes are inevitable if these institutions are to survive. In the institutional theatres attracting the practitioners and copying the methods of smaller experimental theatre has been seen as one solution to attracting new audiences. The most recent and most compelling evidence of the reversal of the traditional order of things was the National Theatre's *Transformations* season. Over the summer of 2002 the National Theatre reduced its audience capacity in the Lyttleton Theatre and created a new theatre space in the Circle Lobby entitled The Studio. These new, experimental, spaces were accompanied by a series of works that were put under tight restrictions both in terms of budget and in terms of time for rehearsal. This complete reversal of the working practices of the National Theatre, to emulate the workings of the experimental theatre, seems to be an indication of a change at the heart of theatre making.

This shift in practices seems to have been combined with by a more sophisticated understanding of the importance of new technology as a communication tool. As a result, the 21st century is starting to see some much more interesting negotiated relationships between the large institutional theatres and their audiences using digital technology. The theatres are responding to a perceived desire for smaller more tailored programmes. The two most striking examples of this trend were the National Theatre's *Transformations* season, already mentioned, and the Royal Shakespeare Company's season of little know plays by Shakespeare's contemporaries. In both of these cases these large institutional theatres created an experimental season that was designed to attract new audiences into smaller spaces to create a sense of a more individualised experience. Both of these seasons struck a chord with audiences as they were quickly sold out. A sense of excitement was created about the new work which brought people back to these theatres or brought them for the first time because they seemed to be offering something quite new and different. The programmes offered challenged audience expectations and perceptions (see Figure 1). It can be seen, therefore, that significant shifts in working practice are beginning to take place that try to balance audience demand with a desire to challenge and inspire.

By contrast, mainstream commercial London theatre continues along very conservative customer-driven lines. Commercial theatres in the West End have for some time looked to film and television stars to prolong their current working practices, however, this is a short term answer to the fundamental problem of an

Figure 1.
The Island Princess (Fletcher). l-r: Shelley Conn (Panura), Sasha Behar (Quisara - The Island Princess). RSC/Swan Theatre, Stratford-upon-Avon. 02/07/02. ©Donald Cooper.

aging audience for conventional approaches to theatre and audiences. The commercial theatre, it appears, is travelling in the direction of the commercial film industry, towards increasingly formalised saleable products for a transient disloyal audience. The work of the Disney Corporation, and in fact the involvement of this conglomerate in live theatre, is the most obvious example of this cross-media marketing strategy.

Shakespeare's Globe Theatre: a new model for the future?

The significant exception to this trend is Shakespeare's Globe Theatre at Bankside which offers a productive new way of working. While Shakespeare's Globe functions entirely without subsidy it has one of the country's healthiest records at the box office and one of the most innovative artistic programmes, as well as extensive education and outreach programmes. The theatrical programme concentrates on combining contemporary ideas about performance with a spirit of enquiry into theatrical practices in Shakespeare's day (see Figure 2). Mark Rylance, the Globe's Artistic Director says of the theatre:

> ...here you are aware of how much Shakespeare kept the audience in mind and what a difference there is when an actor is able to do that too... It looks like a heritage site on the outside, while inside it is the most experimental space in British Theatre.
> (Rylance in Shenton 2002 55)

The audience for this programme is extended considerably by the educational practices of this theatre. The education programme relies heavily on commercial sponsorship for funding and on digital technology for delivery and communication with audience members around the world. Patrick Spottiswoode, Director of Education at the Globe, says:

> At our playhouse, students and teachers meet and work with people who revive words and help them play. Actors, directors, musicians, voice and movement coaches, designers and fight directors, who rouse words into lively action, share their craft and passion with over 50,000 students and teachers every year.
> (Spottiswoode 2003 1)

The Globelink site, a private site which an individual can gain access to for a small fee, offers a range of specialised services and resources for audiences at a distance. Perhaps the most innovative of these is the *Adopt an actor* scheme, a programme that allows international students to follow an "actor's experience as they create a role from the first day of the rehearsal to the final performance in the Globe" (Globe 2003 20). The communication which is set up between actor and student is not one way. The students are invited to offer suggestions to the actor which the actor may use in rehearsal. For example, in one case an actor asked the students to put forward ideas about a hobby her character could have. The suggestion, collecting stamps, was incorporated into the rehearsal process but was eventually dropped when the performance moved on stage because the stamps became invisible to the audience. The result of this exchange engaged the students in the very real process of change that takes place in rehearsal where often ideas, even very good ideas, have to be jettisoned for practical reasons.

Figure 2. *Antony and Cleopatra* (Shakespeare). Paul Shelley (Antony), Mark Rylance (Cleopatra). Shakespeare's Globe, Bankside, London SE1. 29/07/99. ©Donald Cooper.

The Globe's webpage (http://www.shakespeares-globe.org) and private Globelink space connect absolutely with the theatre's live performances, workshops, lectures and theatre tours. The notion of "practice-led resources for the study of Shakespeare" (Globe 2003 20) is combined with an audience driven interactive experience. The model of offering a range of choices for purchase creates a very democratic interaction with the audience which reflects both the theatre's ethos and the strengths of digital technology. The fact that a commercial consumer model has been implemented in a creative way shows that the technology in no way determines the nature of the exchange between the audience and the theatre. Shakespeare's Globe has shown an innovative approach both to its theatrical practices and to its audience interaction before, during and after the performance. This innovative model, which offers a real alternative to the commercial West End, is worthy of emulation.

The National Theatre and the RSC on the web

It is perhaps the strength of identity and purpose that Shakespeare's Globe possesses which makes it strategy so successful. The National Theatre and the Royal Shakespeare Company have both begun to develop increasingly extensive web presences but they show the tension of trying to follow the leads of both the commercial world and the educational world simultaneously. I suggest that the nature of the materials available shows a great deal about the self-images of these theatres. The National Theatre has a quite extensive Education section on their webpage (http://www.nt-online.org/). It is, however, directed almost entirely at schools and particularly at teachers in the UK. There is a very extensive list of INSET training days for teachers and a wide range of downloadable teachers work packs. The *Shell Connections* ongoing project which encourages the writing and performance of work by and for students further indicates the emphasis of the Education programme. The National Theatre quite clearly sees itself in the role of national centre of excellence in both theatre performance and theatre education. The thrust of the education programme is geared towards supporting teachers working within the National Curriculum in the UK.

The Royal Shakespeare Company, by contrast, has a very limited amount of Education material on its website and seems to be struggling rather more with its own identity (http://www.rsc.org.uk). The webpage includes a relatively new exhibition space which has been funded by the National Lottery's *New Opportunity Fund*. The audience for this resource, which includes play specific and thematic approaches to the repertoire, seems very general. As a result, the resources appear unfocused and slightly out of place on a site which is otherwise almost entirely commercial in its aims. The Royal Shakespeare Company's position is made rather more complex by its already existing relationships with both an educational establishment, the Shakespeare Institute, and an archive that holds its materials, the Shakespeare Centre Library. The RSC must therefore be more careful when expanding into these areas not to offend existing partners.

New Artistic Directors have recently taken up their posts in both of these large institutional theatres and it is possible that significant changes may soon be seen in this area. What is clear is that a diffuse and uneven approach to audience relationships through an unfocused web presence can harm as much as help these large institutional theatres. The Globe Theatre model of an education and outreach programme that is fully integrated with the theatrical season provides an example of what is possible when form and content online are successfully united.

Three experiments in collaboration between education and theatre practice

While the Globe Theatre has managed to develop an innovative approach with the support of corporate and individual donations I suggest that innovation is also possible through collaboration between theatre practitioners and educational and other publicly funded establishments. It is my hope that in the future agreements can be reached between the large institutional theatres and the universities, libraries and archives to facilitate publicly oriented, rather than commercial, approaches to the development of long term projects for the study of theatrical production. The commercial world has produced a range of materials in this field of varying quality and longevity. While there are a great many people who might be interested in theatrical history creating materials for this market has not proven to be a viable business venture. I would argue that the preservation of British theatrical history should not be left to market forces and can only be seriously addressed through the collaborative efforts of existing publicly funded institutions in the UK. I offer up three examples of collaborative experiments in this field that follow this model.

King's College/Globe Theatre MA Programme

The Globe Theatre again leads the way in terms of the collaborative relationship they have set up with King's College London to offer a unique MA programme in Shakespeare Studies entitled *Text and Playhouse*. For this programme the Globe theatre have hired a fulltime academic member of staff. The aim of this programme is to combine the performance resources of the Globe theatre with the academic resources of King's College. Students take half of their course work at the Globe and half of it at King's College where they are taught by two Arden editors.

The programme is designed to take advantage of the resources available 'both intellectual and physical' as well as the close proximity of the two institutions. During the one year course students are introduced to the processes of scholarly editing and to original printing practices. They are also encouraged to test their theories about the text in practical ways on the Globe stage. This programme marries together the academic practice of scholarly editing with practical experimentation in a way that has, on a number of occasions, had an impact on the editions published by the course conveners. This example shows the way in which the creative combination of expertise in education and theatre practice can create a unique educational experience for post graduate students.

CalArts Center for New Theater, California Institute for the Arts

The work at the Globe brings together academic editorial practices with theatre practices. The second example I would like to look at brings together theatrical experimentation with theatre training at the university level. This example comes from the United States. The newly created Center for New Theater is the professional production wing of the California Institute for the Arts. Susan Solt, the Dean of the School of Theater at CalArts and the Producer/Artistic Director of the Center for New Theater explains the new company's mission in the following way:

> *The CalArts Center for New Theater (CNT) is the professional producing arm of the California Institute for the Arts School of Theater. Its mission is to provide a framework for the development of theatrical performance work of top caliber with an emphasis on alternative theater and cross-disciplinary work, embracing the diversity of viewpoints that exemplifies the forward-thinking aesthetic of the Institute.*
> *(Solt 2002)*

The first project mounted by this company in June 2002 was an all female production of *King Lear*. This production involved a promenade performance of the play staged in a disused brewery next to a Los Angeles freeway. The production used film and television conventions and employed a wide range of technologies which confronted the audience's expectations.

The use of live feed video and complex sound and lighting innovations resulted in a performance that filled the senses. The all female cast of faculty members from CalArts were ably supported by a thematic and rhythmic chorus made up of student actors. Played against a background of industrial waste and modern technology this production made its audience profoundly uncomfortable on a number of levels. Performed just a few miles from Hollywood the production played on the media driven expectations of its audience (Figure 3).

As a piece of avant-garde theatre this production had a startling and resonant impact on its audience. The power and personality of the actor in the central role combined with the powerfully and uncompromisingly cold interpretation of the director resulted in a production of the play that used a range of technologies to discuss with its audience their own relationship with new technology. The fact that this production was invited to the Centre Nationale Dramatique de Bourgogne's Dijon Frictions Festival 2003 shows how experiments of this kind, stemming from an educational environment, can have an impact on the wider theatre community both nationally and internationally.

The Centre of Multimedia Performance History, Royal Holloway University of London

The final example I put forward is my own research work which has involved the creation of large digital databases of material to contextualise Shakespearean performance history. While the first two examples marry university teaching with theatre practice, my own work draws on the work of theatre professionals to create new kinds of theatrical archives. The *Cambridge King Lear CD-ROM: Text and Performance Archive*, for which I was co-editor, and the *Designing Shakespeare* audio visual archive, for which I was the principle investigator, each have taken a different approach to the use of digital technology for the support of performance history. The *Cambridge Lear CD* replicates many of the strategies of book publication and is fixed in form. The *Designing Shakespeare* project, by contrast, develops a body of information to facilitate a visual approach to the study of Shakespeare in performance and is presented in a flexible way on the web. Through both of these projects it has been my aim to raise questions about the use of new technology for research in this field.

The *Lear CD* attempts to give general access to textual and performance materials which have formerly been available only to private scholars working in restricted reading rooms which are geographically spread across North America, Australia and the United Kingdom. By, on the one hand, drawing together the primary materials of concern for two related, but largely unconnected disciplines and on the other hand opening those materials to a wider range of people through digital technology, an opportunity for new kinds of teaching and research to develop has been created.

The aim of the second project, *Designing Shakespeare: an audio-visual archive 1960-2000*, has been to create and collect performance-based materials which have not formerly been available and which are focused on the temporal and spatial aspects of theatre production. The archive is made up of four databases that include: production credits for all productions of Shakespeare in London and Stratford from 1960-2000; pictures of these productions in performance; interviews with designers and 3D models of the theatre spaces where Shakespeare is most often performed. This collection of four databases is housed by the Performing Arts Data Service at the University of Glasgow and is freely accessible on the web anywhere in the world (http://www.pads.ahds.ac.uk). The objective of this project has been to open a collaborative approach to theatre research at the university level to a public audience. The National Theatre and the RSC, I would argue, would be well served by allowing their archival work to be seen within a wider context of Shakespearean performance history. My own publicly available research work has tried to give the work of the National Theatre and the Royal Shakespeare Company that kind of historical and artistic context.

These three examples of new working practices show, then, that digital technology creates the possibility of new kinds of audience relationship both in the theatre and between the theatre, education and the general public. The institutional

Figure 3.
Fran Bennett as
Lear in the 2002
CalArts Center
for New Theater
production of
King Lear
directed by
Travis Preston.
Photo by Steven
A. Gunther.

theatres face a real challenge in defining themselves within this new realm. New programming initiatives, like the *Transformations* season at the National and the *Shakespeare's Contemporaries* season at the RSC are indications of change. However, it will be the work of the new Artistic Directors in both of these theatres that will potentially begin the shift in working practices in the theatre that the cultural movements of the digital age demand. The Globe Theatre, I suggest, leads the way in terms of creating an egalitarian and engaging experience for audiences at home and abroad which is commercially funded but not only commercial in its aims. My own archival research work and the performance work of the Center for New Theater at CalArts and the King's/Globe MA show ways in which collaboration between theatre artists and educational establishments can support experimentation with new technologies and new working practices.

The future and publicly funded theatres

The National Theatre is already beginning to develop its policy in interesting ways. Nicholas Hytner, the new Artistic Director, announced at the beginning of his tenure a six month season of plays in the Olivier Theatre where two thirds of the tickets were just ten pounds. This, he says, took a significant accommodation in working practice by the directors:

I've been working with my team over the last months not only on our choice of repertoire, but also on new ways of presenting it. In particular, we're determined to make the National available to new audiences and to make it possible for our devotees to come more often... I've asked a group of the English-speaking world's leading actors, directors and designers to collaborate on a bold new way of producing these plays which has made it possible to rethink our ticket prices. The result will be to reveal the Olivier's bare bones and to establish it as one of the most exhilarating theatre spaces in the world.
(Hytner 2003)

Hytner's commitment to make the National "the most public of all London's theatres" (Hytner 2003) is admirable. This idea is supported by the fact that during the recent conflict with Iraq the National hosted a series of early evening gatherings entitled *Collateral Damage* in which actors, writers, comedians and activists took to the stage to share their views about the war in Iraq. This forum combined with Hytner's production of *Henry V*, which featured an overt critique of the embedded journalism of the war, show a real sense of direction and purpose (Figure 4). Hytner's vision of the National as a gathering place and a place of public debate I think very much reflects our current view of the role of the theatre as social discursive space, a view which has been influenced by the democratising medium of the web. The Royal Shakespeare Company has yet to make a significant change of policy or direction of this kind. The plans of the former Artistic Director, which were aimed at popularising the work of the Company in a very commercial way, were universally unpopular. It remains to be seen what position the new Artistic Director will take but it appears that a shying away from this commercialising trend is on the cards.

It seems unlikely that a move away from increased interaction with the audience for any of these theatres is possible. The audience increasingly demand an openness they have come to expect through digital technology. We expect to arrive at the theatre prepared in every sense. At first that expectation was satisfied by the ability to buy tickets online and find information about routes to the theatre and parking availability. Increasingly, however, audiences are keen to learn more about the theatrical process and the current debates raging in theatre criticism and research. Increasingly both the institutional theatres and the universities are being asked to be transparent about their working practices and to engage directly with the public that funds them. The depth and breadth of information required to facilitate this level of transparency, plus the technology needed to create it in digital form may well force the large institutional theatres to enter partnerships either with commercial companies or with educational establishments. Similarly experiments with new technologies in terms of creative practices might productively be developed in or with educational institutions that are not primarily driven by a profit motive.

Conclusions

When creative thinking and new technologies are combined they offer up all sorts of possibilities for change. Some power structures in the theatre are already shifting and I suggest that many more changes will come in the future. Creative

Figure 4.
Henry V
(Shakespeare.
Director:
Nicholas Hytner).
Adrian Lester
(Henry V).
Olivier Theatre /
National Theatre,
London SE1.
13/05/03.
©Donald Cooper.

collaboration, I argue, could bring about a rich and complex future for the theatre and for society. In many ways Shakespeare's Globe Theatre has laid down the gauntlet to the National Theatre and the Royal Shakespeare Company. When it was established the Globe was created as a working theatre, as an educational establishment and as an exhibition space. As a result, specialised staff has been hired and the space has been custom built to accommodate all of these activities. The seasonal nature of the theatre season provides an opportunity for the education programme to have equal status and audience reach. The result is a space that is active every hour of every day throughout the year. The remit of the National and the RSC also seem to be expanding with new public forums and an ever increasing web presence as new aims. The National has focused its developments in education at the schools level and has made a concerted effort to make the theatre more accessible to a wider public through its debates and pricing policies. The RSC has ventured into the area of archival work through its online exhibition, however, I would suggest that this move seems less well integrated into an articulated new direction than the initiatives seen at the National. The RSC may well be hindered

165

in its progress in this area by complex longstanding relationships which exist with educational establishments and its own archive but also by the expectations of its longstanding audiences.

New technologies offer these theatres the opportunity to define more fully their ongoing relationship with their audiences, but the strength of online communication, and the expectations that stem from it, is the two-way nature of the communication developed. The National and the RSC it would appear are attempting to be more open in their approach to audiences. What remains to be seen is whether these theatres will listen and heed what is said by their publics. Can a new more democratic relationship be established between these theatres and an increasingly educated and demanding public? I think it can and it must if the theatre is to remain a vital part of our society. I suggest that these theatres would do well to look to education for models of collaboration that might help them to offer additional services and support for their audiences. The coming together of the rigors of professional practice and of academic research, I suggest, would provide the richest possible experience for future audiences. Shakespeare's Globe Theatre indicates how successful the marrying of these two approaches can be. The answer to the initial question posed must be a tentative yes. It would appear that the National and the RSC are beginning to take the opportunity of reinventing themselves through the use of digital technology quite seriously indeed. If they succeed it will be a success we might all participate in as well as benefit from.

Note

1   In her preface to the second edition of *Theatre audiences: a theory of production and peception* (1997 Routlege, London) Susan Bennett writes: "It is my hope that this new edition will, like its predecessor, provoke studies of specific theatre audiences in specific historical moments, and that the questions asked here will prove interesting ones for others to pursue." While I do not quote her directly Susan Bennett's spirit of enquiry has influences this chapter immensely.

References

Globe (2003) *The Globe Education brochure.* January–December 2003.

Green, R. (2003) Beauty and the Beast. *KDHX Theatre Review.* www.kdhx.org/reviews/ beauty_and_the_beast.html

Hytner, N. (2003) *Welcome to the National Theatre.* Spring 2003 season brochure. National Theatre, London.

Shenton, M. (2002) Smart Ass: Classical theatre superstar Mark Rylance tells Mark Shenton how Shakespeare's Globe lets him be intimate with audiences. *Sunday Express* August 4 2002, pp. 54–55.

Solt, S. (2002) Center for New Theater at CalArts: Statement from the Producer/Director. Part of the *Press pack for King Lear.* CalArts, Los Angeles.

Spottiswoode, P. (2003) Word Play. *The Globe Education brochure.* January–December 2003.

13 New visions in performance

Gavin Carver and Colin Beardon

In the introduction we proposed that, in addition to the individual and creative interest of each chapter and project taken in isolation, a number of themes could be seen to emerge in the book, many of them present in varying degrees in each of the pieces. While it is worth reiterating again that we do not propose that this volume covers the full range of performance work using new technologies, and we do not seek to homogenise these projects simply by dint of their using new technology (and thus being represented in this volume,) there is nonetheless much to be gained from pulling together some (elements) of the emerging praxis.

Virtuality

In its early stages, any new technology will be understood primarily as technique. That is to say, the developer will communicate the intended use of particular gadgets to artists who then incorporates these into their traditional practice, or maybe exploits them in some new experimental pieces of work. As the techniques become more numerous and artistic experience grows, so a more sophisticated relationship between the technology and art form develops. This is the stage we now appear to be at and, hence, it is no longer appropriate to analyse the chapters in this book in terms of the digital techniques that they employ (motion capture, Internet, complex algorithms, etc.)

   Rather, we need to move to a more conceptual level so that we can talk about the technology in terms of models of use. To do this we suggest using the term 'virtuality' as a way of referencing a set of interconnected concerns and manifestations of the new technologies described. This (often overworked) term still has some currency, particularly if we avoid seeing it as a simple homogenous concept and admit that it has several variations (as Burke & Stein make clear in their chapter, and as the reader will see in the explications offered in many others).

   We put forward a three-fold distinction within 'the virtual' in order to describe the various uses of digital technologies within this volume. In doing so we are acutely aware that the neat distinctions that we may make with words inevitably get blurred when we come to consider their application to the messy business of practice—the sharp edges begin to crumble and overlap and the lack of clear boundaries may make us wonder. Nevertheless, there are real differences in the uses of digital technology between, say, Carson and Petersen, or between Dove and Harris and it is important that we find a language to articulate this.

The virtual as stored up potential

Pierre Lévy (1997) has long pointed out that the virtual is not opposed to the real, rather the virtual has a material reality just like everything else. What distinguishes it, in Levy's view, is that it contains at one moment a plethora of possibilities, any one of which can be realised and made actual. This is the view of the virtual as an algorithm or formula that can generate actual examples at will. It is virtual in the sense that it is not itself a specific actual or realised event, and there is a sense in which such virtuality is lost once it is manifest as performance.

One can see several examples of this approach to virtuality within this volume. Sometimes it is a major part of the process, as in Harris' use of algorithms to visualise the appearance of clothing in motion, or Callesen et al's use of algorithms to program the V-actors; the virtuality implicit in the system is made real in moments of improvised performance. It can also play more of a background role, as with Nitsche & Thomas' use of a powerful games engine to bring together the different elements in the recreation of a scene. In other works, this mode of virtuality is less to the forefront, but is present working alongside other modes (see below). For example, while Burke & Stein stress the communicative function, they still rely upon algorithms to transform the audience responses into a database, and then to use this to generate some action (output).

Such virtuality is not, of course, aesthetically or intellectually neutral, a point raised by Povall who uses pre-written motion capture software though he is acutely aware of it as a "pre-cogniscent" environment. The extent of acceptance, or questioning, or re-engineering of software written by someone else, often for some other purpose, is a theme that runs through the contributions.

The virtual as overcoming space and time

There is no doubt that networking, and particularly the Internet, has transformed our use of a computer from a solitary to a public act and that this is a very different kind of use from the algorithmic potential described above. This is virtual in the sense that it enables us to communicate across gaps of space and time. It makes it possible for people (that is, actors and audience, if such a distinction is maintained) to experience a performance while in different places, or at different times, and it also makes it possible for people to communicate with one another, remotely, while making or watching a performance.

Several authors place this communicative role central to their process. Carson, for example, considers the ways in which conventional mainstream theatres have in the past, and could in the future, use the Internet to establish new relationships with their audiences. Both Dove and Burke & Stein explore the potential of using networks (both local and public) to engage audiences in some participative way in performances with a view to changing the power relationships that surround the theatre.

In these cases the 'digital' may seem almost accidental: it happens to be a convenient medium for the transmission of data over communication networks, though it has the added advantage that the data is in the same format as required by the other two modes of virtuality, enabling the kind of mixing-and-matching that we have already noted.

The virtual as transformation and metaphor

One of the reasons that the digital is so powerful is that it provides a common bedrock for so many media forms, and this ability to represent text, images, sounds, etc. opens up many possibilities of transference from one media object to another. The third form of virtuality exploits the convergence of media; for example, where aspects of a performance are captured and the data is then processed in order to generate some kind of metaphorical representation of the original. This is virtual in the sense of "What if *this* were to behave like *that*?", the ability to apply an imaginative, metaphorical transformation. Although a distinct aspect itself this clearly connects to the notion of virtuality as stored-up potential.

One of the major techniques currently used to enable this has been motion capture (using the term broadly to refer to a range of technical processes), where physical properties of a performance are captured and this is then used as a data set from which actions relating to some other object are generated (via some algorithm). Harris, for example, uses a dancer's motion to generate the movement underlying her animations, Callesen et al. explore a range of such relationships within a formal research setting, while both Coniglio and Gilson-Ellis & Povall place the primacy on artistic expression while still exploiting the same technology.

A variation of this can be found in the approach of Petersen, who is particularly interested in the use of new technologies in order to play with the frameworks of space and time. His description of the work of *Dumb Type* raises a similarity of approach to the account of the work of *Blast Theory* by Dixon. The use of digital storage and playback in order to create asynchronous presentations of the same story is a feature of both performances.

While seemingly different, these two approaches can be related at a higher level. The virtual as metaphor is, in fact, the virtual as exploration of new relationships of space, time and causality, for metaphors and similes are, in essence, ways of transposing causal effects from one object to another. Harris makes this clear in her work, for she removes from the presentation the body and objects that have an effect on the movement of garments; once the cause is removed the effect is decontextualised, and oddly draws more attention to the cause than had it been visibly present.

The one chapter that stands outside this categorisation, and hence has yet to be mentioned here, is Smart's analysis of McGregor's latest work. This is because the claim is made that the virtual has itself become the subject matter of the art work and its technical role in supporting the work has become secondary. Interestingly, Gilson-Ellis & Povall adopt a similar approach, deliberately downgrading the role of the technology as technique and allowing it to contribute more "as a whisper", and Harris also seeks to beguile the audience, the technological process being rendered invisible.

Perhaps, then, this second stage in which the technology presents itself in conceptual form, as virtuality rather than as technique, already contains the seeds of its own demise, and that the 'new visions' of our title already point beyond their own horizon.

Finally we might observe that while we have discussed the use of virtual systems to create metaphors and transformations, to translate information into imaginative manifestations, the virtual itself is a metaphor, or perhaps more accurately a 'thematic device'. The deployment of the signs of the virtual (Internet, webs, matrixes, icons, data-streams, etc.) provides a semiotic field in which artists evoke notions of non-linearity, flux, confusion, non-reality. Popular and fanciful manifestations of virtual worlds in film, novels and television, complete with an attempt at physicalising the abstract space of data (what does the web look like from the inside?) provide an arena in which, generally, the organic and independent 'realness' of humanity can be tested. Ironically of course these stories are often told in very linear and analogue media. This mythology of virtuality must have an affect on our readings of any work that engages with this field.

The body

The body is frequently regarded as one of the more problematic entities of contemporary performance and media practice, it provides the site of engagement between the individual and the wider body politic. It is a central feature of the projects cited here where the relationship between (and the space between) the corporeal and the technological provides the primary focus.

The confluence of science, cultural criticism and mass media have alerted us to the subjective, constructed and fragmentary nature of the body, and by extension the problematic relationship between the constructed body and identity, the processes referred to by Foucault as 'the technology of the self'. The construction and control of the body, and the ensuing questions, projections and formulations of identity have become a central feature of contemporary art and performance, not least brought about by technological processes of mediation. The ease with which new media may capture, transform, recirculate and condition the body have particularly sharpened the focus of performances that use such mediating processes, but the question remains open as to whether these processes may be perceived as democratic and liberating or totalitarian and oppressive. Furthermore the facility to record, edit, analyse and replay the actions of the body have problematised the location of the body in a linear flow of time (no longer a stable reference), and questioned the authenticity of performance as a unique moment of liveness. In this sense even the apparent materiality of the body is virtual (and, after Lévy, the virtual is also material).

Popular representations of the body promulgated by advertising, mass media and other technological and cultural mediations suppose that the way we display ourselves and perceive ourselves is culturally (and thus politically) constructed. Furthermore the ability of surveillance technologies to track us, record us, and store us undermines the autonomy and privacy of the individual self. We are subject both to mediated construction and technological scrutiny. This state of affairs is particularly worked out in *GroupSound* described by Callesen et al. where through a combination of visual / aural enticement and surveillance the system may coerce the participant into particular physical responses. While still exercising free will it is clear that the system is the referee (or coach) of the behaviour. In *GroupSound* a simple but sinister set of rules may be read as attesting to the totalitarian aspects of technology's relationship to the body. Indeed here the body is not simply that of the individual, but may be read also as the social or community body. Callesen illustrates this further by indicating that the participants in these systems are not fully aware of who they are in relation to the system, their identity and function must, at least in part, be defined by their relationship with the technological context. In this example the 'system' appears to coerce the participants. However, similar processes result in a situation where the individual body may control a rage of environmental factors by a simple movement, a whim almost. Gilson-Ellis & Povall and Coniglio particularly describe such situations. The fine line between the system providing an extension of the individual and autonomous body, and it acting as dictator of the body's responses, neatly epitomises the contested territory between virtual technologies (particularly the Internet) as democratic liberator and as sinister Big Brother. The empowerment of the individual is no less tyrannical perhaps, but places control with the corporeal rather than digital manipulator.

While this appears to suggest a juxtaposition of the materiality of the body with the abstraction of virtual technologies, the body itself may be regarded as a construct of discourses, of political and cultural operations. The body may be seen as virtual, as a metaphor and site of various potential articulations. One might perceive this relationship to be a dialogue between the virtuality of the constructed body and the virtuality of mediating processes, both as un/real as each other. The relationships are rarely straightforward. Generally the separation between the body and the virtual or technological domains are in a state of flux, and the boundaries between each are shifting, extending each into the other: hierarchies are confused or subverted (as described by Smart), bodies are multiplied or augmented and they extend themselves into virtual worlds as much as the technological extends itself into the body. In many projects the boundary of the skin is not sacred, and it is perhaps, as Broadhurst (2001) has suggested, in the liminal space between body and technology that new creative potentials are to be found. In *Nemesis* technological prosthetics extend the body, in *Kinematograf* the body is duplicated and left in virtual worlds, in *Liquid Views* the observer's body is reflected in the system to become the subject of the piece, in *Chameleons 4: the doors of serenity* the live performer is a cyborg, the dancers for Troika Ranch have movement detectors attached to their bodies, and Harris' projects require that the machine removes (ingests) the live bodies completely. The majority of projects described require that the body be digitised, its movements and enunciations are captured and coded for algorithmic processing. For Burke & Stein this coding applies to the communal audience body whose demographics are captured for inclusion in the piece. Their individual and communal identities (based upon their addresses, demographics, photographs, etc.) are directly included within the fictional world, this is not simply an inclusion of their semblance, but of very potent symbols of identity which allows the theatrical presentation to be truly here and now.

It is notable that a number of the projects are principally engaged with physical performance rather than performance of dialogue, articulated ideas, character, and plots; and even where the latter features (Nitsche & Thomas, Burke & Stein, and to a certain extent Dove and Gilson-Ellis & Povall) the representation and construction of the individual and communal body is a significant feature. While the slight emphasis on dance and physical performance might suggest that the fluidity of the moving organic body is well suited to digital interaction, a number of papers (Burke & Stein particularly) indicate that such fluidity may also be found in the construction and performance of dramatic literature. Coniglio, artistic co-director of Troika Ranch, proposes that the interest is, at least in part, located in the dialectic between the organic (ever changing, never perfect) emanations of the body and the repeatable, predictable sameness of the digital media. The poetry of the encounter is thus partly metaphysical but also expedient, the dancing artist receives immediate feedback from their actions, and thus is provided with an improvisatory framework that has predictable (because they are electronic) and unpredictable (because of performer choice) results. The live movement or vocal articulations are extended into larger scenographic gestures. For Gilson-Ellis & Povall the body extends itself into the environments by means of remotely located tracking technology, a fluid and less violently intrusive version of the cyber-punk notion of 'jacking-in'. In *The technology of the real* Smart describes a very physical manifestation of the extension of the body by and into technology as dancers wear hi-tech prosthetics; here the movements are not just augmented or developed upon by technology, but the body is seen as technological. While real-time interactivity is not as dominant in the work analysed by Smart as it is for some others, the dialogue between the organic and technological is central, as the status and meaning of the body in space and time is problematised by its technological and virtual framework. The digital twitches of the avatars in Nitsche & Thomas's *Casablanca* are perhaps the most idiosyncratic manifestation of the extension of the organic into the technological. In several projects (those of half/angel, Troika Ranch, Callesen et al, Harris and Dove) the relationship between the corporeal and the technological seems quite ludic and this playful and poetic dialectic appears central to the work.

The boundaries between the technological and corporeal are slippery. In a predictable and physical world the body might be a (privileged) constant, in a virtual world where time and space are in flux, more often the body, in its relationship to its identity, form and context, is equally unpredictable. The view of the body as virtual is particularly astute and articulated when performances multiply the body, creating doppelgangers or doubles.

Dixon, particularly, reflects on the nature of these doubles and finds in digital systems an ability to deal with the complex representations of the body / identify / self, and thus able to operate in a conceptual framework that may draw on psychoanalysis as well as anthropology. The virtual double, Dixon argues, may play one or several roles in relation to the self: reflection, alter-ego, spiritual manifestation and manipulable mannequin. The manifestations of the double are seen in a number of projects cited in the book. In *Kinematograf* (Petersen) the performer leaves doubles of herself in virtual worlds as she searches for a reality, and in Dove's *Spectropia* the live performers (who control the virtual worlds) appear on screen, interpreting or mediating between the virtual and the real, a technological spirit guide. In Dove's *Artificial Changelings* the character Zilith, a computer hacker, is described as "manifesting as a polyvocal chorus". In many of these pieces therefore the performer / characters split themselves, both into "other representations/facets" as reported by Dixon, ghosts or traces and an articulation of the dichotomy between the virtual and the flesh. In this latter we may perhaps perceive an articulation of the 'problem' of the body in a media-filled culture, in navigating constructed realities we create empty semblances of ourselves, giving up our body (or at least its image) to mediated worlds. In *Dancing Shadows*, described by Callesen at al. the double1 is exactly that, an apparent shadow, only form, no identity.

Once the body is captured or created, divorced from its original identity, it offers the potential to be controlled or inhabited, remotely, by others or, conversely, it offers the opportunity for participants to take on different bodies. Dove finds that the real body of the viewer starts to mimic the projected virtual body, thereby controlling the performed actions. Harris speaks of the potential for her performances to allow a male dancer to animate a female form and Nitsche & Thomas allowed participants to take on the voices and characters of Bogart and Bergman.

Time and Liveness

The reshaping of time has a particular resonance for performance where the ephemerality of the moment and the fixed duration of the event are significant defining factors. A recurring theme in this volume is the restructuring of narrative and the ability of the audience or participants to reorder events. Dove, particularly, illustrates this possibility in *Artificial Changelings*, in which not only does the movement of the viewer control the order of events (the flow of time), but the piece presents characters from different 'times'—Zilith from the present and Arthusa from the past. In Blast Theory's *10 Backwards* (reported by Dixon) a processed, projected video relay allows a screen to show action that has just taken place flowing backwards and forwards at different speeds, the live performer mimicking actions shown on the screen that she herself has just performed. *Memorandum* by *Dumb Type* (Petersen) similarly stretches and compresses the passage of stage time via video projection. For these and many of the other projects cited the narrative does not obey the fixed linearity normally associated with the passage of time (only it does, because the performance starts and ends), events are reordered by reader or performer, outcomes are not fixed, meanings are assembled not only around cause and effect, but also through association, ideas, histories; a radical denunciation of political inevitability[1]. Or is it? For in some cases the cycle is repeated over and over, linear time at least provides escape. Such flexibility and plasticity of time not only problematises the stability of cause and effect (and the place of the individual body within a known and stable structure) but also questions a central ontological element of performance[2]; the liveness (and immediate death) of the moment. This, coupled with the ability to create virtual (technologically projected) echoes and reflections of people not actually present creates an ironic paradox in the taxonomy of performance; it is both more live (as it is not fixed, open to change) and less live (as it may not be corporeally present).

The debate over the primacy of liveness in the hierarchy of performance qualities is implicit through the chapters in the book, and reflects a real critical problem facing contemporary performance in a media culture. In the face of cinema and television (and now, perhaps the Internet and computer games) it is arguably the liveness of theatre that is its distinguishing feature. Liveness seems to suggest four things.

Aspects of the event are open for change, and those changes may be witnessed in the moment, this may be as subtle as an extended pause not present on a previous showing, or as dramatic as a trapeze artist falling during a move.

These changes may come about through interaction (as much as by individual choice or 'fate'), not just between performers, but also between the audience and the performers, thus emphasising our participation in the here and nowness of the event.

Human agents are actually present during the performance, I say 'agent' rather than performer, and 'actually' rather than visibly to account for puppet theatre (a form perhaps closer to many projects in this book than conventional western theatre). Nonetheless the point is that there is flesh and blood present, work is seeing to be done in the here and now, and the work produces sweat and spit.

Once past the moment cannot ever be recaptured.

It is arguable of course that the elevated of the status of liveness is an atavistic sentimentality, a construct to preserve 'theatre' in a world of recorded and widely distributed performance (Copeland 1990). It is also worth noting of course that the economies of distribution enable filmed performance to have far higher budgets than theatre—bigger sets, fabulous sound, great actors, the live theatre has a corner to fight perhaps.

It is axiomatic that the inclusion of telematic technologies in performance destabilises the definition of genre or form, for while liveness (immediacy, irreproducibility) may be a central feature of performance, digital technologies tend to mean the inclusion of projected, non-live, potentially repeatable elements, which in some work vies for hierarchical supremacy with the live performer. In *Liquid Views* (described by Dixon) the sole observer, projected into the space, becomes the performer; in the projects of Callesen et al. the events are animated by 'participants', a performer-audience hybrid; in Dove's *Artificial Changelings* the performer is projected as if in a film; in Nitsche's *Casablanca* the performers are either remote, live puppeteers or video game avatars; and in Harris' pieces the performer has been removed entirely, having 'done their bit' by feeding movements to the motion capture system. The pieces represent a number of stops on a continuum from more or less traditional live performance (in that identifiable performers present themselves in front of a fairly passive audience) such as *Nemesis*, Troika Ranch, or half/angel, through interactive live 'events', to work that is more or less video installation. Far from the suggestions that this is evidence of performance being subsumed by mass media technologies the range of projects seems to provide evidence that theatrical and live performances are reconstituting themselves, embracing related forms of play, ritual and film. While of course projection has been used in live performance throughout the 20th century (for example in the works of scenographers such as Piscator and Svoboda) the digital projector / processor combination has made image manipulation and display accessible, powerful and affordable. The relative cheapness and ease of use of digital projectors and video cameras can reclaim recorded (or digitally constructed) moving images from the industrial, mass market context of the film industry, redeploying them in more intimate, experimental contexts, where the context of production allows for the integration and interaction of the live and recorded image. Theatrical performance has the facility to use new media alongside human agents (performer or audience), who can be seen 'in the moment' to interact with it, to control and be controlled. This offers a powerful critique and challenge to the hegemony of media culture. The languages of film and television can be subverted and redeployed outside of the hegemony of mass / popular culture, and the human agent (whether on stage or at a terminal, actor or performer) may be seen to have a direct effect. It may be, as Auslander has observed in a different context, that the ability of theatre to take on, subvert and redeploy mediating / reproductive technologies may be more politically potent than its refusal to do so in its attempt to preserve an atavistic ontology of liveness (Auslander 1997).

Of course on a different level many of the works in this book are extremely 'live' since the parameters for interaction and change are far greater and more profound than those allowed within a lot of performance practice. The viewer is highly active in the construction and presentation of the narrative, the scenography and space respond to the performer directly and in the moment, the combined outcome of which is far less easy to reproduce than conventional theatre. The inclusion of new technologies may on the one hand problematise the role of live, corporeal presence in the hierarchy of performance (this in-itself is in an important aspect of the performances) but equally may dramatise interactive possibilities for audience and performer more clearly.

At the opposite extreme in this book, Harris presents a project in which the live dancer is absent, not simply absent because the viewed product is a digitisation of her dance, but absent because she has been removed from it entirely, all that remains are the dancing clothes (digitally created). Where for McGregor the dancers extend themselves into the technological realm, for Harris the dancer is entirely transported into it. Although Harris clearly articulates her desire to develop a live version of the project (and she clearly spells out the technological problems), here the non-liveness of the piece sharpens more clearly the problems of presence / identity in performance with new technology. Harris makes work that she feels (as do the editors) is 'performance', but since the performer is wholly (in the end) digital, its stage has to be the screen.

Interactivity

Intimately connected to the extension of the body into the technological domain, and the plasticity of time, is the potential for interactivity in the performances; one of the most cited, and arguably most significant features of these works. Any digital system provides a means of processing information, extremely quickly, allowing for a real time response. The algorithmic articulation of virtuality facilitates the almost immediate playing out of one of a range of possible events, facilitated by interaction (conscious or otherwise) between performers, space and audience / participant. In many of the projects described the input is derived from the performed movement of a participant (Coniglio, Callesen et al., Dove, Gilson-Ellis & Povall, Harris), while in other cases the driving input is a more direct interface with the computer (akin to a computer / video game), such as the case in Nitsche & Thomas. For Burke & Stein, and to a certain extent Carson, some of the input is remote from the event and requires a web interface. While these systems operate in a variety of different ways the end result is to effect an interaction between the body (of either observer/participant or performer) and the environment. The input need not be restricted to physical movement, the voice of the performer may be read and processed, or the input may be in the form of typed text. However gathered, once the data has been input into the system a software engine generates a responsive output; Burke & Stein propose this input / output structure should be considered a central element in constructing texts for performance with new technologies. The output may be visual, either projected into the performance space or displayed on a monitor, or audio — music, sounds or text. This is the principle behind what half/ angel call 'poetic systems'; it is the interaction between participant and system that creates the poetry and narrative in a number of the projects.

The dialogue between performer and system problematises authority / authorship. The performer may, to a certain extent, be the author of their actions in the moment, but the interactive system (potentially programmed with algorithms hidden from the performer) adapts or plays with the input, creating a new meaning. Similarly the output (sound or image) may have an original composer / author, but the interaction of performer and system / algorithm makes a new work. Richard Povall describes such a system at work in *Spinstren*, arguing that this combinatory system defies closure. At a further extreme the interactive performance of *Casablanca* wonderfully demonstrates the conflicting authority present in intermedial performance; the games mode of the interactive software intrudes into the vocal performance of Bergman and Bogart, who are in turn 'directed' by on-line puppeteers, fancifully the random twitches of the avatars seem to indicate the difficulty in sustaining this three-way tension, or occupation of the body.

While for Coniglio and half/angel the system provides a responsive space in which performers extend or augment their live presence, reaching into the performance environment, for Callesen at al. the system may act as a manipulator or facilitator. In a number of their projects live participants work with(in) the system, each providing input for the other, and the actions of the human participants are prompted by the computer (via sound or image projection) which then responds to those movements as an improvisatory game develops. In these contexts the system can take the appearance of a character, and entity which appears to have a presence. And indeed the space may become a storyteller, weaving a narrative in which a participant may become actively involved. In essence, Callesen et al's system is similar to other interactive models, but in the main their interest is with the processual and improvisatory possibilities offered to unrehearsed participants, a game-playing scenario in which the relationship between the system (the V-actor) and the participant drives the narrative events.

For Harris the notion of interactivity is more subtle, more virtual perhaps. While there is no actual possibility of interaction in that the performance is entirely preset, she suggests that the absence of a performer (she produces work that displays garments moving without inhabitants) allows the audience to mentally place themselves within the piece, substituting themselves for the absent body.

The role of the active viewer is a central feature of Dove's chapter in which she describes a project in which the system offers the viewer / participant the opportunity to structure the narrative of what is effectively a film-story. The interface is as invisible as possible, based on the movement of the viewer's body, it translates the natural body responses and movements into mechanisms that edit the film. The system facilitates a shared authorship of the artefact and clearly illustrates the way in which interactive technologies may problematise fixed and unilateral narratives, the identity of the active viewer aids in the construction of the identities of the characters. However, in the field of performance, this in itself is a problematic position, for as Burke & Stein illustrate the act of performance is bounded by a progression through linear time; a performance starts, proceeds forwards and finally stops. The audience do not have the ability to roll time backwards to review other possibilities, any one reading of a performance text is inevitably linear. Thus for Burke & Stein effective interaction requires that the audience are aware of their presence within the piece, not via the somewhat abstract notions of interactive democracy, but by incorporating hard information about their identities in the performance text. This inclusion of the audience community in the text is a central feature of their interactive performance making, and its facility provides for a new way of making texts.

For Carson the viewer is active in a rather different way. Her chapter proposes that it is not just the interactivity in the moment of performance that is an essential step forward, but an dialogue between the consumers and producers of performance and theatre provided by digital information systems, particularly the Internet. Potentially, this reverses a cultural economy that was based upon the producers being distanced guardians of what they supposed was wanted / needed, and allows the wider population a role in the process of creative production. However this can also simply be a cosmetic exercise that uses the tricks of targeted and inclusive marketing to create the appearance of responsiveness. But if studies of popular culture have told us anything it is that when people have the means to articulate their cultural desires (however subcultural), then institutional producers of culture will pick it up and recycle it; cultural hegemonies are not clear-cut[3]. The Internet and digital technologies provide ample access to the dissemination of these desires.

Space

We have considered how these works extend the body into space, and make space an active agent in the pieces (Callesen's notion of the anthropomorphic space), similarly the disruption to the flow of linear time makes space plastic as the 'now' is constantly shifting, so is the 'here'. The Internet and remote communications extend the space beyond the architectural confines of the place of performance, performers and participants need not share the same space at the same time. Projection, whether real-time interactive or not, facilitates the presentation of spaces and people not actually present on stage. While this is true of traditional film and transparencies as much as it is of digital images, the digital process allows for a manipulation of these images, or a complete construction of them. Digital image production and processing allows the fusion of the real and the unreal, it enables the superimposition of a captured live event into an impossible environment, or conversely, the projection of a hyper-real environment into the 'artificial' world of performance. Real performers in real spaces (who are fictional characters in poetic spaces) combine with spaces and characters who are not corporeally present but may seem more 'real' or more abstract. This 'conceptual dichotomy', as Smart terms it, provides one of the anchors of Random Dance Company's piece *Nemesis* (Smart) in which the levels and conflicts of 'the real' assist in confusing and clarifying their comprehension. Working mostly with space (real and unreal) the viewer's perception of reality, scale and context is continually challenged, while the live bodies of the performers appear both within and outside of these constructed spaces. Petersen also reviews projects which "suspend the concepts of space and time as fixed definitions", arguing that the use of telematic systems, which enable the dislocation and distortion of space and time, requires audience and theatre maker to develop a new language of theatre.

Across such examples, questions about the reality of theatre space and (projected) virtual space are sharpened, but what is notable is the range of expression given to this projected / other space. Gilson-Ellis & Povall tend to use unmediated representations of the real world (images of a meadow for example) whereas McGregor presents an apparently real image of a room which is 'worked-upon' throughout the piece, its mutation drawing attention to the 'problem' of the real in performance. *Kinematograf*, analysed by Petersen, uses clearly 'artificial' worlds, deploying the tropes the 'virtual'. Similarly Nitsche's interactive *Casablanca* presents its location designs without feeling the need to mask its artificiality, using a visual style and communication codes derived from a video game aesthetic: the designs for the virtual environment for *Casablanca* consciously refer to *Metropolis* more than Morocco. Perhaps a feature that comes out most clearly when reading across these chapters is the range of aesthetic modes and strategies employed within the telematic systems, and what may need addressing is how we, as readers of the work, can reconcile and decode the range of approaches? Arguably theatre depends upon representational systems that are open for interpretation on a number of levels: a chair on stage is automatically semiotically rich for the very fact that it stands in front of an audience on a stage where the very nature of the frame implies its meaning may slip from iconic to indexical to symbolic; it may simply be a 'chair on stage' (never just a chair) a throne, 'the state', a horse and so on. It is possible that the looser, less 'convincing' visual language of computer games may support the complex semiotic utility needed by the stage, whereas photo realistic images may lack the possible flexibility or slippage, they have too much gravity.

As much as virtual systems may offer multiple views of time they may also offer multiple perspectives on space. Multiple cameras, with real time processing may present simultaneously, multiple perspectives, potentially instantaneously distorted, replayed, zoomed in, etc. One is offered a point of view that is far from neutral and distanced, the world may be seen simultaneously in micro and macro, live and mediated. Furthermore, where present, the effect of multiple perspectives (cameras replaying the stage from various viewpoints for example), or the effect of performance time replayed and modified, draws more attention to the presence, immediacy and multiplicity of performance. Iain Mackintosh (1993) has celebrated the multiple perspectives offered by playhouses of the Restoration, but in essence, for any one viewer, the point of view was singular, and perspective staging was only really 'right' from a single point in the middle of the auditorium. Telematic (surveillance) technologies destabilise the authoritative and singular point of view, and offer all members of the audience differing / multiple views—none need be privileged. At the same time this multiplicity of view points reaffirms presence and liveness, the event is three dimensional and multifaceted.

Magic

It is interesting how many times a spiritual or supernatural metaphor (or indeed text) appears in the articles. For Dixon this is partly a reclamation of sympathetic magic; digital and telematic technologies can recreate doubles, and through the double one may create an effect on the original; virtual systems may create an unnatural totemic domain. Of course, the situation is more complex since the performance as a whole is providing a double to life, the opportunity offered by digital systems offers a further doubling. One may perceive how digital worlds are similar to 'other realms', levels of existence that cannot always be accessed by normal mortals, and Dove, for example, provides a guide to give participants access. The possibilities of the 'virtual' to create a counterpart that is more abstract and ephemeral than the physical performer, and far more so than the 'real' for which the performer, in turn, doubles. The virtual is magic in the sense that it provides a range of possibilities, of transformations, which may be instantaneously activated and accessed by a shamanic performer.

For Gilson-Ellis & Povall the technology provides a tool in which a responsive space may offer a poetic and magical augmentation of the work, and the very first sentences of their chapter allude to the unearthly / spiritual possibilities. Words appear in the air, lost songs are sought, and the story revolves around folkloric mythology. Dove also peppers her article with references to haunting and to ghosts, and Callesen et al. based a project around an interactive / participatory ghost story. Harris digitally recreates the movement of clothing, based upon the captured motion of a dancer, but in replay removes the body: unlike Dixon's double, Harris draws attention to questions of identity through the absence of a body/form and provides a very magical / unearthly image in the process: in her terms it is beguiling and seductive. It is not hard to see why interactive and telematic systems produce effects that may be perceived to be magical or spiritual, but it is interesting that that this is an outcome of an entirely predictable and material process. Inscribed within a number of these projects is a clearly anti-technological aesthetic; while the creative and processual possibilities are celebrated, they are put to work in projects that resonate with more archetypal and mythological concerns. That is not to say these concerns evade the problematic postmodern questions posted by digital technologies, issues of subjectivity, social constructions, commodification, subjugation (and resistance) are all present, but often approached elliptically and poetically.

Extending beyond the performance

Digital systems facilitate the extension of the performance beyond the prescribed time and place of the central performance event. It is well understood that the pre-

and post- phases of performance (gathering and dispersal) play an important role in constructing the frame of the performance, and contribute to the decoding of the central performance event[4]. The paraphernalia that surround a performance, and not simply the mechanics of promotion, are a meaningful part of the activity. Carson argues that the growth of digital communications technology (principally of course the Internet) has made gathering and dispersal an increasingly complex and significant part of the event. More significantly, she sees this relationship as becoming more audience centred; the relationship, at best, is not simply a dissemination of marketing material, but a set of inclusive (interactive) practices which fosters an increasingly democratic relationship between (potential) audience and the work.

For Burke & Stein this pre-performance participation is also a significant feature of the work. The audience first engages with the performance (not simply its advertising rhetoric) via the Internet, and in doing so provide certain personal details that will later be used directly in the performance text, constructing an event that responds to a specific community. For them the real potential of digital technology is not in its contribution to the spectacular, but in its ability to provide material with which to customise the narrative; upon which they develop the case for a new thinking about writing for performance.

The virtual worlds constructed in the abstract and interactive domain of cyberspace may be rich locations for performance or meta-performance, facilitating apparently democratic, interactive exchanges and obfuscating the relationship between truth and fiction (how do we know truth on the Net?); but also harvesting detailed information about the participants. Behind the use of information gathering technologies there lurks the spectre of surveillance and totalitarianism; the democratic may be seductive and illusory. This ambivalence between freedom and subjugation is threaded through new media, and surfaces frequently in its critiques. In Callesen et al's projects the system takes on a persona, it uses surveillance technology to monitor the activities of its participant / subjects and then coerces them into particular patterns or routines. Stelarc's work, described in this volume by Dixon, is more explicit still in its critique of the relationship between 'democratic' technology, power and the body. These pieces do not only extend the performance frame well beyond the event itself (geographically and temporally), but they do so in a contested domain which may be both democratic and totalitarian, fictional and authentic, real and abstract. Blast Theory's performance *Kidnap* (2000) was viewed via the Internet, through which its audience witnessed the imprisonment and treatment of two captives[3]. This performance depended upon surveillance and communications technologies and clearly made the audience voyeurs. The fact that it existed in the same frame as internet news channels, sub -cultural and counter-cultural websites, and entertainment media (such as Big Brother) draws attention to problematic slippage in the media—authority: entertainment; revolution; control?

Nitsche & Thomas extend their *Casablanca* project beyond an identified performance event in a number of ways. Firstly they draw upon digital media's propensity for intertextuality, present of course in performances without new media, but the assemble / edit process particularly facilitates intertextual references. Their performance brought together the vocal performances of Bergman and

Bogart playing a scene from *Casablanca* (romantic melodrama) using the technologies and aesthetics of video games, an aesthetic that is quite brazenly 'digital' and a form that is conspicuously not romantic, though perhaps erotic. The romantic couple are embodied in avatars that are derived from commodified gaming stereotypes. Furthermore the performance was played over a network. Performer / puppeteers manipulated the characters and the point of view / camera in real time, reacting to the developing scene from their own terminals. Although one assumes that for the participants the experience was live, the performance itself was entirely virtual, and could, conceivably, be witnessed by any number of individuals from their own terminal. Here the performance event seems largely dissipated, or extended away from a significant moment; its content and scenography references external sources (albeit with bespoke designs) and its playing has no conventional performance framework. Read from a conventional performance-critical position, this problematises the act of performance, whereas read from the perspective of interactive visual media it does not.

Destabilisation

We have illustrated a range of key themes present across the chapters in this book (albeit with different emphases and manifestations): interactivity through a variety of models; the construction / presentation of virtual worlds and 'double' performers; the problem of liveness and presence. All of these necessarily destabilise the performance in that the narrative is not clearly fixed, the reality of space and presence is held open to question, multiple perspectives are present and made explicit, individuals are represented as dichotomised elements, and time and action are not necessarily linear. Even seemingly straightforward relationships between performer and audience are subverted, and theatrical hierarchies are unbalanced in that space, sound and (projected) image are as dominant as the performer, who in some cases is not present at all. While several writers make an explicit point of this destabilisation (Smart, Dove, Petersen, Dixon, Burke & Stein) its manifestations are present across all the work. But these destabilisations do not seek to subvert the form of the performance as such, since their emphasis is on liveness—ephemerality and contingency. One could argue that this reaffirms the centrality of the immediate in the theatrical experience, and where the use of technology may traditionally be seen to mitigate against spontaneity / immediacy (as manifested in spectacular but inert hi-tech scenographies), here the technology enhances the liveness. This is made further apparent when the body of the present and live performer is juxtaposed with their double, or augmented through technology.

In 2000, *The Times* previewed an interactive performance due to take place at the University of Kent and, understandably but mistakenly, worried that projections in performance would seek to illustrate the text, making the imaginative poetry of the play redundant in what they suggested might be an overly simplistic imaging of the words. Far from simply illustrating texts, the examples offered in this book align themselves more closely with what Birringer describes as the "[mobilisation of] audiovisual techniques of abstraction against the no longer clearly defined / confined phenomenological reality of the stage" (Birringer 1991 175). In a similar vein, writing in 1996 in response to a project by Mark Reaney at the University of Kansas, Delbert Unruh suggested that these abstract, interactive virtual worlds free the theatre from the tedious mechanics of concrete staging and, more importantly perhaps, from the representational verisimilitude prevalent in Western theatre, he suggested the new scenography provided new explorations in the representation of time and space (Unruh 1996). The chapters in this volume certainly provide us with those new dimensions, new visions, and make significant augmentations, not least to the role of the body and the opportunities offered by interactivity.

Notes

[1] These themes are explored by Susan Kozel, pp.257–63 in Goodman and Gay (2000), in which she develops the term 'post-linear' to describe works that consciously "plays with the 'space, time, and context of narration" (p.259).

[2] A position perhaps most strongly articulated by Peggy Phelan (1993) in the definition of what she calls the 'ontology of performance', for Phelan it is the liveness, and ultimately death of performance that defines it.

[3] John Fisk (1989) illustrates this particularly will in his essay on the cultural and commodity role of jeans.

[4] Discussed by, amongst many others, Bennett 1990, Schechner 1977.

References

Auslander, P. (1997) Ontology vs. history: making distinctions between the live and the mediatized. *Proceedings 3rd Annual Performance Studies Conference*, Atlanta, Georgia, April 1997. http://webcast.gatech.edu/papers/arch/Auslander.html

Bennet, S. (1990) *Theatre audiences: a theory of production and reception.* Routledge, London.

Birringer, J. (1991) *Theatre, theory, postmodernism.* Indiana University Press, Bloomington and Indianapolis.

Broadhurst, S. (2001) Intereaction, reaction and performance: the human body tracking project. *Body, Space, Technology* 1(2), (electronic journal at: http://www.brunel.ac.uk/depts/pfa/bstjournal/1nol2/journal1no2.htm accessed 11th September 2003)

Copeland, R. (1990) The presence of mediation. *The Drama Review* XXXIV(4) 42.

Kozel, S Introduction to Part 8: Post-linearity and gendered performance practice. In Goodman, L. with de Gay, J. *The Routledge reader in politics and performance.* Routledge, London.

Lévy, P. (1997) Welcome to virtuality. *Digital Creativity* 8(1) 3–10. Reprinted in Beardon, C. and Malmborg, L. (eds.) (2002) *Digital Creativity: a reader.* Swerts & Zeitlinger, Lisse, The Netherlands, pp. 7–17.

Mackintosh, I. (1993) *Actor, architecture, audience.* Routledge, London.

Phelan. P. (1993) *Unmarked: the politics of performance.* Routledge, London.

Schechner, R. (1977) *Essays on performance theory 1970–1976.* Drama Book specialists, New York.

Unruh, D. (1996) Virtual reality in the theatre: new questions about time and space. *TD&T* Winter 1996 44–46.

Notes on authors

Colin Beardon is Professor of Computer Graphic Design at the University of Waikato in New Zealand. He produced the 'Visual Assistant' software for rapid visualisation of theatrical scenes and has been involved in many projects that combine his skills in computer science and experience in the creative art and design community. He is co-editor of the journal *Digital Creativity*.

Jeff Burke is a researcher and visiting faculty at the School of Theater, Film and Television, University of California, Los Angeles where he is a co-architect of *The Iliad Project*. His newest work is a collaboration with molecular biologists and geneticists, *In Silico*, which will premiere at the Fowler Museum of Cultural History in Fall of 2003.

Jørgen Callesen received his MA in Information and Media Science from Århus University in 1995, and worked as a planner for theoretical studies at Space Invaders in Copenhagen, until 1997. He studied puppetry at the theatre school 'Ernst Busch', Berlin 1998–99 as part of PhD studies at Aarhus University. He is currently Lecturer in Multimedia Design and Production at Aarhus University.

Christie Carson is a Lecturer in the Department of Drama and Theatre at Royal Holloway, University of London and Director of the Centre of Multimedia Performance History. She is the co-editor of *The Cambridge King Lear CD-ROM: Text and Performance Archive* and recently completed *Designing Shakespeare: an audio visual database, 1960–2000*. In March 2003 Dr. Carson was seconded as the Project Officer for C & IT in the English Subject Centre at Royal Holloway.

Gavin Carver is Senior Lecturer in Drama and Theatre Studies at the University of Kent where he teaches in areas of theatre design and visual performance. He has published on the use of computer modelling in scene design. He is the co-founder of the Kent Interactive Digital Design Studio, a research group investigating the role of digital technology in performance practice and research.

Mark Coniglio is a composer/artist who focuses on the creation of works that combine music, dance, theatre and interactive media. With choreographer Dawn Stoppiello he is co-founder of Troika Ranch (www.troikaranch.org), a New York City based performance group committed to creating works that integrate dance, theatre and interactive digital media.

Steve Dixon is a senior lecturer in Media Performance at the University of Salford, UK. He is artistic director of the multimedia performance company The Chameleons Group and co-director (with Barry Smith) of the Digital Performance Archive (art.ntu.ac.uk/dpa). He has published widely on topics including cybertheory, theatre, gender, film and performance studies.

Toni Dove is a US-based artist/independent producer who works primarily with electronic media, including virtual reality, interactive video installations, performance and DVD ROMs that engage viewers in responsive and immersive narrative environments. Her work has been presented in the United States, Europe and Canada as well as in print and on radio and television. (www.tonidove.com)

Jools Gilson Ellis is co-Artistic Director of half/angel, a Cork-based writer, choreographer, performer and installation artist. Her work is focussed on developing provocative poetic spaces in different performance platforms. She is a Lecturer in English at University College, Cork, Ireland.

Jane Harris is a Senior Research Fellow at Central St. Martins College of Art and Design and a visiting lecturer at Goldsmiths College, London. Her work has received awards from The Arts Council of England (2002), The Arts and Humanities Research Board (2001), The Arts Council and Channel 4 (1996). Jane's work has been exhibited and presented extensively in the UK and internationally.

Marika Kajo received a Master of Science in Computer Technology at Lund, Institute of Technology, Sweden in 1989. Her early work centred on rehabilitation robotics, before entering multimedia and flexible learning for people with special needs. She later continued studies in Theatre/Drama at the University of Trondheim, and the University of Copenhagen, where she also took part in actors training.

Katrine Nilsen was awarded a degree of Production & Stage Design from Denmark's Design School in Copenhagen in 1998. Since then she has been working as a designer for exhibitions, film, theatre and dance performances in Denmark and Sweden. Since 2001 she has been part of the research project Performance Animation Toolbox, at the Interactive Institute in Malmö, Sweden, developing new effects and methods for virtual stage design.

Michael Nitsche has been involved in practical research projects in the UK (with Sony and National Film and Television School) addressing dramatic virtual environments. He wrote the script for a commercial game (*Zanzarah*, Funatics Game Development, THQ). He will submit his PhD on 'Virtual story spaces' at the Digital Studios at the University of Cambridge in 2003 and will then take up a postdoctoral research position.

Kjell Yngve Petersen is a stage director and performer. He is Artistic director of Boxiganga. He was trained in the old traditions of advanced formal body theatre, in the tradition of Danish Odin Theatre, Japanese No, Italian Commedia del' Arte and French Abstract Mime, by Etienne deCroix, guided by the performer Maria Lexa. He works on creating augmented and hyper-reality performances and installations, informed and experienced through the body, with special interest in real-time generative artwork where the audience is performing the artwork. He is currently a PhD student at the Planetary Collegium, University of Plymouth. (www.boxiganga.dk)

Richard Povall is co-Artistic Director of half/angel, a Devon-based digital artist, composer and researcher whose work is concerned with adapting and extending technologies to build evocative spaces. He teaches occasionally at Dartington College of Arts, UK.

Jackie Smart is a Senior Lecturer in Drama, Faculty of Arts and Social Sciences, Kingston University, UK. She has written on dance for the camera, on contemporary dance and devised theatre (most recently, 'On Narrative: an interview with Tim Etchells', in *Frakcija*), and on performing arts pedagogy. She also engages in practice-as-research with her physical theatre company, Quisling.

Jared Stein is a playwright and artist-in-residence at the HyperMedia Studio, University of California, Los Angeles, USA. He is the other architect of *The Iliad Project* and also the resident playwright of the Rhodopski Dramatichen Theatre of Smolyan, Bulgaria.

Maureen Thomas is a dramatist, screenwriter and interactive media director. Since 1998 she has been Creative Director of Cambridge University Moving Inage Studio (CUMIS). She is Associate Professor in Narrativity andInteractivity at the Norwegian Film School, and from 1993-1998 was a HeadTutor at the National Film & Television School, UK. Her creative research focuses on the spatial organisation of narrative, non-linear and aleatoric hypermovie, and navigable interactive story/drama in real and virtual space.

Bernard Walsh is Deputy Senior Tutor, Visual Arts Department, Goldsmiths College, London University and a lecturer on the MA Fine Art course at Central Saint Martins College of Art & Design, London.

Index

L

M

R

S

INNOVATIONS IN ART AND DESIGN
ISSN 1570-7393

1. Digital Creativity. A Reader
   Edited by Colin Beardon and Lone Malmborg
      2002. ISBN 90 265 1939 7 (hardback)

2. New Visions in Performance. The Impact of Digital Technologies
   Edited by Gavin Carver and Colin Beardon
      2004. ISBN 90 265 1966 4 (hardback)